D0290608

CATACLYSMS

CATACLYSMS

A New Geology for the Twenty-First Century

MICHAEL R. RAMPINO

COLUMBIA UNIVERSITY PRESS

New York

Columbia University Press
Publishers Since 1893
New York   Chichester, West Sussex
cup.columbia.edu

Copyright © 2017 Columbia University Press
All rights reserved

Library of Congress Cataloging-in-Publication Data
Names: Rampino, Michael R., author.
Title: Cataclysms : a new geology for the twenty-first century /
Michael R. Rampino.
Description: New York : Columbia University Press, 2017. | Includes
bibliographical references and index.
Identifiers: LCCN 2016058574 (print) | LCCN 2017025755 (ebook) |
ISBN 9780231544870 (e-book) | ISBN 9780231177801 (cloth : alk. paper)
Subjects: LCSH: Natural disasters—Environmental aspects. | Comets—Collisions
with Earth. | Volcanic eruptions. | Extinction (Biology)
Classification: LCC GB5014 (ebook) | LCC GB5014 .R355 2017 (print) |
DDC 551—dc23
LC record available at https://lccn.loc.gov/2016058574

∞

Columbia University Press books are printed on permanent and
durable acid-free paper.
Printed in the United States of America

COVER DESIGN: Milenda Nan Ok Lee

COVER IMAGE: Johan Swanepoel © Shutterstock

Contents

Acknowledgments

I want to thank my colleagues Ken Caldeira, Yoram Eshet, Andreas Prokoph, Stephen Self, Richard Stothers, and Tyler Volk, who coauthored papers with me on the wide-ranging subject matter of the book and who provided a sounding board for my ideas. Ernest Gilman, professor of English at New York University, read all the chapters and made many cogent comments that helped me refine the text. Barbara Richardson helped with making the writing more reader friendly. Tracy Pursell went through the manuscript and made a number of improvements in the style. For discussions and information, I thank Asfar Abbas, Jonathan Adams, Walter Alvarez, David Brin, Phillipe Claeys, Rodolfo Coccioni, Chuck Drake, Doug Erwin, Nick Eyles, Rhodes Fairbridge, Al Fischer, Simone Galeotti, Bill Glen, Richard Grieve, Paul Hebner, Alan Hildebrand, Marty Hoffert, Bill Holser, Bob Jastrow, Dennis Kent, David King, Christian Koeberl, David Kring, John Matese, Florentin Maurrasse, Dewey McLean, Yasumori Miura, Simonetta Monechi, Sandro Montanari, Richard Muller, Adriana Ocampo, Chuck Officer, Paul Olsen, Carl Orth, Lucille Petruny, Kevin Pope, Lisa Randall, Dave Raup, David Schwartzman, Jack Sepkoski, Shu-zhong Shen, Gene Shoemaker, Jan Smit, Maureen Steiner, Jessica Whiteside, and George Williams. I know I am leaving some people out; this book had a long gestation period. Jennifer Deutscher prepared many of the figures. The library facilities of New York University and the Goddard Institute for Space Studies were invaluable. Patrick Fitzgerald at Columbia University Press saw the project through from the start and provided much-needed encouragement. I would like to thank the other great people at Columbia University Press for their help in the editing and production of this book: Ryan Groendyk, Kathryn Jorge, Milenda Lee, and Irene Pavitt.

CATACLYSMS

• • • •

Introduction

A philosopher should once in his life doubt every thing he had
been taught.
RENÉ DESCARTES, *MEDITATIONS ON FIRST PHILOSOPHY*

Sir Charles Lyell (1797–1875) was one of the most influential geologists of
his day. His *Principles of Geology*, first published in three hefty volumes
between 1830 and 1833, went through 11 subsequent editions over the next
45 years. The *Principles* would stand as holy writ for geology well into the
twentieth century. For Lyell, the underlying principle of geology on which
any true understanding of Earth's history must be based is that change
occurs only very gradually over long periods of time, involving geologic
processes that we can observe working today. He abhorred then-current
theories of catastrophic geology, in which

> we hear of sudden and violent revolutions of the globe, of the instanta-
> neous elevation of mountain chains, of paroxysms of volcanic energy. . . .
> We are told of general catastrophes and a succession of deluges, of the
> alternation of periods of repose and disorder, of the refrigeration of the
> globe . . . and other hypotheses, in which we see the ancient spirit of
> speculation revived, and a desire to cut, rather than patiently to untie,
> the Gordian knot.

"Never," Lyell tells us, "was there a dogma more calculated to foster indo-
lence, and to blunt the keen edge of curiosity, than this assumption of the
discordance between the former and the existing causes of change."

If catastrophism offered only a hypothetical history of random and unpredictable cataclysms of unknown nature, then there would be little or nothing for geologists to do. There would be no way of discovering any pattern or formulating any rational theory to explain this catastrophic Earth history. Such was the charge leveled by Lyell against two proponents of this pernicious cataclysmic "dogma": the contemporary French catastrophist Georges Cuvier and his colleagues, and, more broadly, those Christian fundamentalists for whom Noah's flood was proof positive of a catastrophic and recent Earth history.

Rejecting as mere "speculation" the claim that Earth was prone to sudden upheavals, so that change was "discordant" rather than harmonious, Lyell preferred a slow and rational picture of Earth's evolution, and his pervasive influence in geology soon replaced catastrophism with sweeping and restrictive gradualist notions that held sway for 150 years. Recently, however, catastrophism has returned to geology, opening up new frontiers with potentially revolutionary ideas. I believe that we are caught in the middle of a paradigm shift in Earth sciences—a shift so enormous and exciting that it may surpass plate tectonics in its importance to our understanding of Earth history. When I was an undergraduate, the continents were still stationary. Earth's history was paced by the gradual rise and fall of the sea, with occasional periods of mountain building. By the time I arrived in graduate school at Columbia, the continents were drifting as the plates moved across the face of the planet. I propose that we are now in a similar situation with regard to the role of cataclysms in terrestrial geology and the evolution of life.

I found mass extinctions of life, especially of the dinosaurs, fascinating even before the announcement in 1980 in the journal *Science* that Walter Alvarez's group at Berkeley had uncovered an iridium anomaly at the time of the Cretaceous/Paleogene mass extinction (formerly the Cretaceous/Tertiary boundary), 66 million years ago. Iridium is a rare metallic element related to platinum. The father-and-son team of Luis and Walter Alvarez discovered the unusual iridium-rich layer at the same level in the rocks where the dinosaurs are known to have disappeared from the fossil record. Could this anomaly be a clue to a mass extinction for which no compelling reason had been given? They attributed the excess iridium to the effects of a giant impact of an iridium-rich asteroid or comet that caused the mass extinction and spread iridium-tainted dust worldwide.

Clearly, the Alvarez team had produced the most sensational discovery in geology in decades. It sent a shock wave through the scientific world, and it had repercussions in all branches of geology, and in astronomy and biology as well. Their findings linked critical events here on Earth with events in the solar system, outside of Earth. It turned out to be the beginnings of what I see as a new Copernican revolution for geology, with the Alvarezes in the role of a new Copernicus. We are not isolated in space—events from outside Earth can affect terrestrial history.

In 1981, I participated in the excitement of the first meeting on the topic, "Large-Body Impacts and Terrestrial Evolution: Geological, Climatological and Biological Implications," held in Snowbird, Utah. For four days, more than 100 scientists from fields ranging from planetary geology to micropaleontology presented arguments concerning the importance of impacts in Earth history. Initially, I have to admit, I was not a supporter of the impact hypothesis. I gave a presentation proposing that the anomalous iridium content of the layer may have come from submarine processes that could dissolve limestone (rocks made of soluble calcium carbonate) and thus concentrate the insoluble noble metals. Such a process is known to produce geologic layers called hardgrounds, marking gaps in the geologic record. This idea did not go over well with the Alvarez team. Alternatively, I was inclined to the idea that the boundary clay layer may be of volcanic origin, having seen many such thin layers of altered volcanic ash while studying global volcanism. Everyone knew that volcanism had been intense in the Late Cretaceous world. I thought that the Alvarez group had jumped too quickly to their conclusion of large-body impact. Apparently, I was still reluctant to throw off my Lyellian bias against extraterrestrial catastrophes.

Shortly thereafter, however, new evidence came to light that changed my opinion. Researchers found that the iridium-rich layer was unique and truly global, as volcanic ash would not be, and that the anomaly appeared in nonmarine sediments in the Raton Basin of New Mexico. This ruled out any possible oceanic concentrating process for the anomalous iridium. These findings convinced me that the catastrophic impact hypothesis was correct. Some of my colleagues criticized me for changing my mind, but in science one should be led by the facts. Fossil evidence showed that the impact occurred at the precise time of the mass extinction in the oceans, and very close to the time of the last of the dinosaurs. By 1983, the question of impact was firmly settled for me by the discovery, in the boundary layer, of grains of the mineral quartz, showing features that can be produced only by hypervelocity shock.

The Alvarezes' discovery caused many researchers, including me, to drop whatever we were doing and to focus on the events at the end of the Cretaceous period. Eventually, scientists found and analyzed more than 350 end-Cretaceous geologic sections from all over the world, on land and at sea, and their study provides a detailed look at the catastrophic dimensions of the impact event. Related effects, such as darkening of the sky from a global dust cloud, wildfires set by the in-fall of hot debris, severe acid rain, massive submarine landslides, and towering tsunamis have now entered mainstream thinking. The "smoking gun" for the impact came with the discovery (or, should I say, rediscovery) in 1991 of the huge Chicxulub impact structure in the Yucatán, and its precise correlation with the time of the mass extinction and impact fallout layer.

Many geologists initially refused to accept the reality of the end-Cretaceous impact event and its effects on life, and some were (and some still are) quite vocal in their disagreement. I think that a large part of the resistance to the impact hypothesis came from the long-standing practice, laid out by Lyell in the early nineteenth century, to avoid extraterrestrial causes for geologic events, and to avoid catastrophes at all costs. Paleontologists especially did not like the intrusion of physicists like Luis Alvarez into their territory. Vertebrate paleontologists who studied dinosaurs denied the impact hypothesis most vehemently. Today, most will admit that an impact occurred, but many refuse to believe that the dinosaurs were done in by a rock from space.

Inquiry has taken many directions since the Alvarez team shook the foundations of twentieth-century geology. The mass extinction at the end of the Cretaceous period is not the only one in the geologic record, and not even the most severe. While scientists around the world were hotly debating the impact story for the end-Cretaceous event, paleontologists presented new evidence that the past 542 million years—the interval with abundant fossils—has been marked by four or five other mass extinctions, as cataclysmic as or even worse than the end-Cretaceous event, along with a number of lesser extinction events.

Could these also have been caused by impacts? Perhaps a general theory of impact-induced extinctions is implicit in these data. The discovery of the end-Cretaceous iridium anomaly set off the search for traces of impact at other times in the geologic past. But searching for impact evidence in the geologic record is a difficult proposition. Iridium and shocked minerals are present in minute quantities, impact layers are very thin, and impacts of large comets (mostly ice) may leave little trace. Nevertheless, as I recently pointed out (based on years of research by a number of geologists), the largest impact craters of the past 250 million years appear at the same time as recognized extinction events, and these craters all seem to be associated with layers of impact debris in the geologic record. This cannot be a mere coincidence. Nonetheless, many geologists cling to the idea that the end-Cretaceous impact was a one-off event, hoping that purely earthbound processes are to blame for the other extinctions.

In 1984, surprising new evidence for an apparent 26 million–year cycle in the timing of the mass extinctions came to light. Since the end-Cretaceous event was already associated with a large impact event, this suggested that all or most of the mass extinctions in the cycle might have resulted from periodic comet or asteroid impacts. The possibility of periodic impact-induced extinctions led to a search for cycles in the record of impact craters on Earth. Richard Stothers of NASA and I jumped into the fray and found a similar cycle, of about 30 million years, when we analyzed worldwide impact crater ages. It seemed, from the evidence, that periodic impacts could be causing periodic extinctions.

To explain the periodicity, scientists presented several astronomical hypotheses involving perturbations of comets in the distant Oort cloud, resulting in comet showers on Earth. This includes my own idea that comets could be perturbed by the cyclic motion of the solar system swinging through the disk-shaped galaxy, and those of others, who have proposed comet disturbances by a small solar companion star named Nemesis, or by an undiscovered Planet X. As might be expected, presentation of these new, conflicting ideas led to some heated discussions among the supporters of the various competing mechanisms. For the galactic model, Stothers and I envisioned that, as the solar system passed through the crowded midplane of the galaxy every 30 million years or so, the concentration of stars and clouds of gas and dust (now including an invisible disk of dark matter, recently proposed by Lisa Randall and colleagues at Harvard) shook the Oort cloud at the fringes of the solar system, sending a barrage of comets toward Earth.

Meanwhile, other geologists pointed out that the end of the Cretaceous was also the time of the catastrophic eruption of the Deccan flood basalts of India, where more than 1 million cubic kilometers (240,000 cubic miles) of lava, covering one-third of the Indian subcontinent, erupted in a relatively brief period. Was cataclysmic volcanism involved in the end-Cretaceous and other disasters? The plot thickened when Stothers and I realized that a number of the giant flood basalt episodes in Earth's past correlated with times of extinction, and also with times when the oceans became stagnant and severely depleted of dissolved oxygen. Several research groups have argued that the environmental effects of such cataclysmic flood basalt eruptions might be severe enough to trigger mass extinctions. So destruction might come from above or below. It has even been suggested that large impacts can in some way trigger increased volcanism.

Back in the 1980s, when I was working for NASA, I noticed that various kinds of geologic events—such as massive volcanism, mountain building, creation of volcanic hotspots, and fluctuations in sea level and climate (all related through plate tectonics)—seemed to be happening on a similar 30 million–year schedule. This had been pointed out by some geologists in the early twentieth century but then largely ignored by the scientific community. It is possible to develop a scenario, based on astrophysical theory, in which dark matter (undetectable except for its gravitational effects), concentrated in clumps near the midplane of the Milky Way galaxy, would be captured by Earth. High concentrations of dark matter particles could lead to their mutual annihilation while inside the planet. This might trigger violent heating of Earth's core, resulting in pulses of geologic activity and volcanism with the same underlying 30 million–year cycle that we see in impacts and mass extinctions. So episodes of Earth's internal activity and cataclysmic volcanic eruptions might also have an extraterrestrial pacemaker.

The Alvarezes' discovery and the studies that followed could mark the beginnings of a new geology for the twenty-first century, a cataclysmic geology that takes into consideration the effects on our planet of the wider solar system and galaxy. Surprisingly, many of the geologic forces that shaped Earth may have their origin in extraterrestrial cycles and in the interaction of Earth with galactic dark matter. This is not good news to some geologists. After one of my talks on the subject, a geologist came up to speak with me, and he was alarmed that, if I were right, the astronomers would take over his science. In reality, the new discoveries of potential links between what goes on in the solar system and the galaxy and events here on Earth promise to weave the sciences of geology and astronomy into a more coherent story.

This book tells the story as I have outlined it here. It begins, in chapter 1, with a focus on the history of geology relevant to the questions of gradualism versus catastrophism in Earth history and, in chapter 2, the establishment of what I call "Lyell's laws," which determined, until quite recently, how we looked at the geologic record. In chapter 3, I consider the revolutionary Alvarez hypothesis, which relates the severe mass extinction of life at the end of the Cretaceous period to the impact of a large asteroid or comet at that time, and the discovery of the large Chicxulub crater in the Yucatán.

Following this, in chapter 4, I look at the record of extinctions published by David Raup and Jack Sepkoski, and the possibility that the extinctions they document are periodic. Chapter 5 then considers the effects of large-body impacts on the environment. In chapter 6, the question of how evolution by natural selection fits into the context of this new catastrophism is explored, with a discussion of the largely unsung early-nineteenth-century horticulturalist Patrick Matthew, who actually proposed a theory of evolution by natural selection before Darwin and who also believed (as Darwin did not) that catastrophes in the history of Earth played a role in the appearance and disappearance of living species. Chapter 7 argues for potential correlations between some mass extinctions and impacts. The greatest extinction, at the end of the Permian period, and its causes are the subject of chapter 8. The very important question of volcanism in the form of massive flood basalt eruptions and their effect on the environment as a potential cause of some mass extinctions is the subject of chapter 9.

Chapter 10 discusses ancient deposits, attributed to glaciation and other causes, that may in reality be the result of impacts. Chapter 11 introduces the galactic oscillation hypothesis for periodic impacts and mass extinctions, including new information regarding dark matter in the galactic disk. Chapter 12 takes up the idea that other important geologic events (for example, volcanism, mountain building, variations of sea level, and climate change) may be periodic as well, related to Earth's motion through the galaxy and its interactions with clumps of mysterious dark matter. In the epilogue, I summarize the situation with respect to the inclusion of catastrophes and cycles into a new cataclysmic geology that could help to bring together the astronomical and geological sciences.

1

• • • •

Catastrophism Versus Gradualism

Looking back dispassionately into the history of geology it is interesting to observe how deeply conservatism has become entrenched.

ALEXANDER DU TOIT, *OUR WANDERING CONTINENTS*

I decided to become a geologist when I was about seven years old. The mother of a schoolmate took us to the American Museum of Natural History. Like Stephen Jay Gould, I was impressed by the amazing dinosaur skeletons on the fourth floor of the museum, but the stunning mineral and gem displays on the first floor also fascinated me. Various rocks and minerals collected from vacant lots and parks near my home in Brooklyn already occupied an entire bookshelf in my room. The museum collection showed me that beautiful mineral specimens occurred all over the globe, which more than kindled my other great desire, ignited by my love of maps and stamp collecting, to travel to faraway places in pursuit of rocks and fossils. Around the same time, my grandfather gave me a book on astronomy, with great pictures of galaxies and nebulae. I was captivated, and a little fearful, contemplating Earth's place in infinite space, but the nighttime sky in New York City was not conducive to astronomical study.

In school, we built volcanoes using glycerol and potassium permanganate. (Such a thing would not be possible in current classrooms; volcanoes are now safely made with vinegar and baking soda.) While other kids got chemistry sets, I got a geology set, with lots of minerals and the apparatus needed to test their various properties. But I ached to know more about geology. In a search for geology books at the local secondhand bookstore, the best I could do at the time was an ancient copy of *The Geological Story Briefly Told* (1875), by James Dwight Dana (1813–1895; figure 1.1). Unknown

FIGURE 1.1 James Dwight Dana (1813–1895), author of *The Geological Story Briefly Told*.

to me, Dana was the preeminent American geologist of his time. His *Manual of Geology* (1863) was a widely used geology textbook in the second half of the nineteenth century, going through several editions.

Dana, a professor of geology at Yale University, came from a long line of missionaries and educators. As a young man, from 1838 to 1842, he explored the Pacific with the Wilkes Expedition, and he published articles on coral reefs and coral islands, in parallel with Charles Darwin's early voyages and publications. Later, in the 1880s, he led the first geological studies of the volcanoes of Hawaii.

I recently went back to my copy of Dana's book (I still have it), which is intended for "the general reader and beginners in the sciences." As I turned its yellowed pages, I was struck by something that I had not fully taken into account during my early readings. Dana's geology had strong theological underpinnings. Time and time again, he mixes descriptions of the workings of geologic processes with theological interpretations. In his view, all the geologic events in the past were guided by God and directed toward one goal: the appearance of humanity.

How had I missed this? I found it hard to believe that Dana was writing in 1875, sixteen years after the publication of Darwin's *Origin of Species*. Surely, the introduction of natural selection should have dispelled the notion of directed evolution. But changes in scientific attitudes do not happen overnight. Handwritten notes by a student who had marked the margins of my book show that geology courses were still using Dana's book until around 1900.

I had assumed that Dana was a world-class observer and scientist, and that the antiquated idea of geologic evolution as part of a miraculous process designed for the creation and benefit of humankind had been banished from the sciences by the time he published his book. Yet here was the famous Dana saying that all the world's geologic processes through countless epochs were designed to make the planet suitable for *Homo sapiens*: "The world by gradual steps reached its present perfected state, suited in every aspect to man's needs and happiness—as much so as his body; and it shows throughout the same Divine purpose, guiding all things toward the one chief end, Man's material and spiritual good."

Even in his later career, in 1885, Dana went so far as to publish *Creation; or, the Biblical Cosmogony in the Light of Modern Science* in an effort to reconcile geologic history with biblical accounts of creation. It is not surprising that Dana also opposed the idea of evolution: "Until the long interval is bridged over by the discovery of intermediate species, it is certainly unsafe to declare that such a line of intermediate species existed, and as unphilosophical as it is unsafe." Dana's writings show that, even into the late nineteenth and early twentieth centuries, some geologists were still reading the rocks with the goal in mind of conciliation between the geologic past and the tenets of Scripture and were still seeing the appearance of humans as the prime objective of millions of years of evolution.

When I was a graduate student at Columbia in the early 1970s, I studied historical geology with the famous Marshall Kay, who was in his sixties at the time. Kay attended university in the 1920s, so his teachers would have been students in the late nineteenth century, when ideas like Dana's were still widespread in the science. Thus the teachers of my teachers were presented at university with a picture of geology very much tied to theological interpretations. The history of Earth was represented as a record of changing sea levels, with slow and steady transgressions and regressions of the seas over the continents. It is no wonder that geology was for a long time caught up with notions of the design of a peaceful world. The truth is that, in some ways, modern geology is really not so far from the geology of the late nineteenth century.

Geology is a relatively new science. It was introduced as a fixed term by Horace-Bénédict de Saussure only in 1779. It grew partly out of "natural theology," as espoused by the theologian William Paley (1743–1805) in his famous essay *Natural Theology, or Evidences of the Existence and Attributes of the Deity Collected from the Appearances of Nature* (1802). His book, which was one of the most published of the nineteenth and twentieth centuries, presents a number of teleological and cosmological arguments for the existence of God. Natural theology was proposed as a way to explain the apparent design of the world and of living things, as part of God's rational plan. It takes its motivation from the far older, overtly theological belief in the legibility of the "book of nature." God reveals himself in Scripture and in the Creation. The book and the world thus offer two parallel texts. The first offers evidence of God's direct intervention in human affairs. But if, in a more "enlightened" age, one comes to regard the Bible as simply allegorical, then it becomes possible nevertheless to find evidence of divine purpose working through the features of Earth in a manner capable of rational study. That there is also supposedly an "order" among living beings, from "lowest" to "highest" (that is, humans), and that we stand at the pinnacle of creation by virtue of having been given the gift of reason and the knowledge of our creator, offers further "evidence" of the same providential intention.

At the time, Christian doctrine held that Earth as we know it was created in six days. In the 1650s, Bishop James Ussher (1581–1656) of Ireland undertook the calculation of the life spans of the prophets in the Bible, and other information, with the goal of figuring out the exact date of Earth's creation. The result—Sunday, October 23, 4004 B.C.E., at 9:00 A.M.—was subsequently inscribed in Bibles as the official Christian date of creation. The implications for the study of Earth were clear: if the world was only 6,000 years old and had been created in six days, then Earth had essentially no deep history, and any geologic changes must necessarily be of a recent and catastrophic nature.

The influential late-eighteenth-century German geologist Abraham Gottlob Werner (1749–1817; figure 1.2) of the Freiberg Mining Academy proposed that the geologic record could be derived from consideration of the expected series of materials rapidly precipitated from a primal world ocean, perhaps during the Great Flood. This explanation was readily compatible

FIGURE 1.2 Abraham Gottlob Werner (1749–1817) of the Freiburg Mining Academy.

with Scripture. Werner used the universal ocean to explain the sequence of rock types—with the densest on the bottom and lighter rocks at the top—that he supposed to exist in all places on Earth in the same order. He explained the crystalline rock granite, which he believed formed the base of rock sequences all over the world, as a chemical precipitate, deposited from the alkaline floodwaters as they first receded. Overlying rocks were the result of the rapid settling of sediments according to their density, all of this taking place very quickly. Werner's theory did not explain how to dispose of the excess water that once covered the world, but no matter.

Werner's idea of rock deposition from a universal sea became known as Neptunism. It is often said that Werner was no field geologist. Had he been, he would have realized that his theoretical rock sequence was no more than that—in many places, granites can be seen to intrude older sedimentary rocks that show signs of intense heating. Furthermore, sediments of various types are found interbedded in no standard order, contrary to what Werner predicted.

These Neptunist views were challenged by the theory propounded by the Scot James Hutton (1726–1797; figure 1.3) of a machinelike, dynamic Earth, driven by subterranean heat, with grand geologic cycles of decay and renovation—representing the continuous interplay between the great forces of uplift and erosion—stretching back to the uncharted beginnings of geologic time. Hutton was a gentleman farmer and businessman. A member of the Scottish Enlightenment, Hutton discussed science and philosophy with the likes of Adam Smith, David Hume, and James Watt.

Hutton's geological ideas were of the "Plutonist" school. He argued from close examination of hand specimens and outcrops that crystalline rocks such as granites and basalts were originally molten material that had forced its way up from deep below and intruded into previously existing rocks or erupted at the surface. He and his followers eventually provided unequivocal evidence for the once molten origin of what are now called igneous rocks—generated at high temperatures in Earth's hot interior.

Moreover, Hutton was a gradualist who envisioned a world in which the results of slow and steady geologic processes such as erosion, deposition, and uplift accumulated over eons to create the major changes in geologic history. These long-term cyclic changes required periods of time much

FIGURE 1.3 James Hutton (1726–1797), author of *Theory of the Earth*.

longer than 6,000 years. Thus Hutton is said to have discovered "deep time"—the countless ages represented in Earth history. His classic observations of outcroppings (figure 1.4), where he could see that marine sediments had been uplifted and deformed, only to be completely eroded and overlain by a newer set of strata, showed that incredibly long periods of time must be involved in geologic history.

Across the English Channel, French geologists and paleontologists of the early nineteenth century, led by Baron Georges Cuvier (1756–1830; figure 1.5) of the Jardin des Plantes in Paris (and often called the "father of vertebrate paleontology"), favored a catastrophist theory of geologic change. Cuvier was a brilliant comparative anatomist. With his colleagues, he carefully studied the fossils and layers of rock in the Paris Basin and reported empirical evidence for episodic catastrophic and sweeping changes. They

FIGURE 1.4 Siccar Point, Scotland. Marine rocks have been deformed, uplifted, and set vertically, and then eroded and covered by more recent marine sediments, in what geologists call an angular unconformity. This sequence of events must represent an enormous passage of time. (Photo by Dave Souza)

found that the geologic record gave evidence of long periods of quiet alternating with brief times marked by the sudden disappearance of fossil species—mass extinctions of life. Cuvier's catastrophist interpretation claimed that the extinctions were sudden and involved unknown, cataclysmic forces. He argued that "we shall seek in vain among the various forces which still operate on the surface of our earth, for causes competent to the production of those revolutions and catastrophes of which its external crust exhibits so many traces."

The mass extinctions, followed by the appearance of new organisms, were commonly interpreted in the context of a "progressional" theory, in which Earth history reflected a movement toward perfection, culminating in humans and the modern world. Cuvier gave no reasons for the repetitive cataclysms, beyond the fact that they were outside the normal realm of geologic processes. The "Preliminary Discourse" section of Cuvier's work was translated into English as *Essay on the Theory of the Earth* by a former

FIGURE 1.5 Georges Cuvier (1756–1830), the "father of vertebrate paleontology."

student of Werner's, Robert Jameson (1774–1854) at the University of Edinburgh. Unfortunately, Jameson's version included his own biblical references to Genesis and the Great Flood, which were not part of Cuvier's original text. So in the English-speaking world, Cuvier was closely associated with biblical catastrophists.

By contrast, Hutton sought to explain the former changes in Earth's crust exclusively by reference to ordinary processes:

Not only are no powers to be employed that are not natural to the globe, no action to be admitted of except those of which we know the principle, and no extraordinary events to be alleged in order to explain a common appearance. . . . Chaos and confusion are not to be introduced into the order of nature, because certain things appear to our partial views as being in some disorder. Nor are we to proceed in feigning causes, when those seem insufficient which occur in our experience.

This was a step in the right direction from theories that proposed biblical or supernatural causes, but Hutton maintained a deistic interpretation of geology. Geologic causes were only secondary causes. His theory was advanced to "bring the operations of the world into the regularity of ends and means, and, by generalizing these regular events, show us the operation of perfect intelligence forming a design." The precepts of natural theology still held strong.

Hutton's *Theory of the Earth* was finally published in 1795, just two years before his death. In it, he laid out the central idea that nature worked in great cycles of destruction and renewal. His theorizing was done with the underlying assumption of a world governed by a balance between land and sea. Continents are first submerged, covered by sediment, then uplifted into mountains, only to be reduced by erosion and again covered by the sea. Areas that were once land became sea and areas that were sea became land, but the world as a whole remains about the same—a steady state—allowing mankind and its animals and plants to survive. The result is a perfectly balanced geologic cycle stretching backward into deep geologic time.

Hutton's theology is apparent in statements such as "If we believe that there is an almighty power, and supreme wisdom employed for sustaining that beautiful system of plants and animals . . . we must certainly conclude, that Earth, on which this system of living things depends has been constructed on principles that are adequate to the end proposed, and procure it a perfection which is our business to explore." For final causes, Hutton appealed to the Creator:

Thus we are led to inquire into the final cause of things, while we investigate an operation of such magnitude and importance, as is that of forming land of sea, and sea of land, of apparently reversing nature, and of destroying that, which is so admirably adapted to its purpose. Was it the work of accident, or was it the intention of that Mind which formed the matter of this globe, which imbued that matter with its active and its passive powers, and which placed it with so much wisdom among a numberless collection of bodies, all moving in a system?

But Hutton did leave the door open to more rapid events, and he stated, "In thus accomplishing a certain end, we are not to limit nature with the uniformity of an equable progression although it is necessary in our computations to proceed upon equalities." Hutton also concluded, "From what has actually been, we have data for concluding with regard to that which is happening thereafter."

So he is saying that study of the geologic record of things past could be used to infer what is to happen in the future. In other words, "the past is the key to the future" or, in the words of Marshall Kay, my former mentor at Columbia, "if something did happen, then it can happen." I call this the first principle of geology. Recently, Hutton's statement that "the past is the key to the future" has been reborn as the motto of the Geological Society of America. I think that this new incarnation is a result of our growing success in reconstructing ancient worlds, in determining the causes of past geologic and evolutionary change, and of the realization that significant environmental changes are happening at the present time (now defined by some geologists as a new geologic epoch, the Anthropocene).

2

Lyell's Laws

No causes whatever have from the earliest time to which we can look back, to the present, ever acted, but those now acting; and that they never acted with different degrees of energy from that which they now exert.

CHARLES LYELL TO RODERICK MURCHISON, 1829

Modern geology derives its core principles from the British geologist Charles Lyell (1797–1875; figure 2.1). Scottish by birth, Lyell had private means, so even though he trained as a lawyer, he was free to pursue his passion for geology. He honeymooned among the rocks in Switzerland and Italy and later published two popular geological travel guides based on his visits to North America. His magnum opus, *Principles of Geology, Being an Attempt to Explain the Former Changes of Earth's Surface, by Reference to Causes Now in Operation*, came out in three volumes from 1830 to 1833 and sold well, making money for its author. Knighthood followed by a baronetcy contributed to his reputation as the most influential geologist of the mid-nineteenth century. Moreover, for a long period, the science of geology has followed three fundamental postulates that can be gleaned from Lyell's text:

1. Geologic change is the product of slow and gradual processes that we can observe today, acting over long periods of time. Lyell mocked the idea that catastrophic changes occurred in Earth's history, and he railed against the geologic catastrophists: "We are told of general catastrophes and a succession of deluges, of the alteration of periods of repose and disorder, of the refrigeration of the globe, of the sudden annihilation of whole races of animals and plants."

2. Geologic forces are intrinsic to Earth. Comets or other extraterrestrial bodies are not to be invoked to explain geologic history. For Lyell,

FIGURE 2.1 Charles Lyell (1797–1875), author of *Principles of Geology*.

astronomical catastrophism "retarded the progress of truth, diverting men from the investigation of the laws of sublunary nature, and inducing them to waste time on the power of comets to drag the waters of the ocean over the land—on the condensation of the vapors of their tails into water, and other matters equally edifying."

3. The geologic record does not contain regular repeating patterns influenced by celestial cycles, as this smacked of predestination to Lyell. Could God allow the stars to influence his orderly world? Lyell ridiculed such theories, proposed by naive astronomers who mistakenly compared the course of the events on our globe with astronomical cycles: "Not only did they consider all sublunary affairs to be under the influence of celestial bodies, but they taught that on Earth, as well as in the heavens, the same identical phenomena recurred again and again in perpetual vicissitude."

Other geologists espoused these ideas before Lyell, but they were codified in his *Principles of Geology* and passed on through generations of textbooks. This is Lyell's legacy, and it has been the fundamental paradigm of the Earth sciences. The motto of geology became "the present is the key to the past"—in other words, one must study the mostly gradual geologic processes now in operation and extend these same slowly acting processes over geologic time to help explain the past. (Note that this is very different from James Hutton's view that "the past is the key to the future.") Even the plate tectonics revolution of 50 years ago did little to change the dominance of these Lyellian views in geology. In fact, the machinelike Earth of plate tectonics is the very model of a slow and orderly system, driven by invisible internal forces.

Although many geologists may not realize it, Lyell's edicts presume that we inhabit a planet designed for human occupancy. This is similar to the views of Hutton and James Dwight Dana. We enjoy a geologic history that represents the unfolding of a calm and orderly process leading to us. Change, when it happens, is always slow and gradual. According to Lyell, this was God's plan for Earth: "In whatever direction we pursue our researches, whether in time or space, we discover everywhere the clear proofs of a Creative Intelligence and of His foresight, wisdom and power." Lyell's picture of quiet gradualism paced by steady internal geologic forces had deep theological roots and did not come from an impartial evaluation of the geologic record. In fact, whenever he was faced with what looked like the record of sudden geologic changes, Lyell was forced to conclude that they were just illusions, the products of an imperfect geologic record. So Lyell produced an influential geological textbook, but one still wrapped up in early-nineteenth-century ideas of natural theology.

Discoveries in the Earth sciences over the past 35 years suggest that Lyell may have been mistaken on all three of his pronouncements. First, geologic changes are not always slow and gradual. Cataclysms have occurred. Second, the forces that govern the planet's biological and geologic evolution may not all be terrestrial, and, third, there may in fact be grand geologic cycles driven by astronomical circumstances. The new discoveries at the forefront of research in geology, astronomy, and related fields are so far from Lyell's laws as to argue for a new geology, a science that rejects much of what Lyell had to say, in order to explain many critical events in Earth's history.

The traditional focus of geology has been primarily inward-directed, and observations about Earth were for a long time limited to the local and small scale. We still tend to see our world in a kind of pre-Copernican perspective, as the center of the universe and yet also strangely divorced from the rest of the cosmos. But we now have evidence that the history and evolution of Earth and life are linked inextricably with the larger universe. The planetary constituents of our solar system were formed within stars and in violent supernovae explosions, the planets condensed from a nebula containing gas and dust surrounding the newly formed sun, and the process of star and planet formation was probably triggered by shock waves from a nearby supernova. Earth accreted in a process of collisions between smaller preplanetary fragments, and it suffered a heavy early bombardment that may have been important in providing the planet with volatiles and organic compounds. For this reason, geology should perhaps be considered a branch of planetary science, which itself is a sub-branch of astrophysics. In its true astronomical context, Earth is part of the larger universe and represents only a minute body in a sea of space.

Astronomy and geology developed independently of each other. By the mid-nineteenth century, the science of geology was moving from studies designed to explain biblical catastrophes to an explanation of the natural world as the gradual workings of familiar forces of nature over enormous periods of time, with Lyell's doctrine being the end point. At the same time, astronomy was moving from descriptions of the movements of the heavenly bodies to hypothesizing about the evolution of Earth, the moon, the solar system, and the universe at large. It wasn't until the middle of the twentieth century that astronomy began to impinge on geology as more and better data were obtained on the moon and planets, and these bodies became suitable for geological studies. The two major geologic processes that affected the planets were found to be impact cratering and volcanism. Planetary geology and planetary astronomy were soon seen to be overlapping fields, together contributing to a new geology of the solar system.

Lyell's reply to the findings of Georges Cuvier and the French school of catastrophists was that the apparent cataclysms and sudden geologic and biological changes resulted from a grossly imperfect geologic record. The sharp geologic boundaries that were seized upon by catastrophists as evidence of sudden and drastic events were merely illusions created by the incompleteness of the rock record. The geologic record resembled a book that could be read, but an imperfect one, with many pages missing because of erosion or nondeposition. Thus, according to Lyell, we cannot believe our own observations, especially when our observations run counter to the orderly "plan of nature" that we are a priori able to deduce.

Lyell's approach is clearly subjective and deistic. Yet many geologists continue to follow Lyell's edict that the geologic record actually records past events poorly. It has been pointed out that Lyell was trained in law and that *Principles of Geology* is really a legal brief. It begins with a particular theological and philosophical view of the geologic record and is constructed as a long argument, with examples specifically chosen to support this view. As an example of a tight a priori argument, its conclusions seem unassailable.

Principles of Geology, as a brief, proved to be extremely effective in the court of scientific opinion. In the end, gradualism carried the day, and Lyell triumphantly proclaimed that "all theories are rejected which involve the assumption of sudden and violent catastrophes and revolutions of the whole Earth, and its inhabitants." One of the most often cited passages from the *Principles* declares, "In our attempt to unravel these difficult questions, we shall adopt a different course, restricting ourselves to the known or possible operations of existing causes; feeling assured that we have not yet exhausted the resources which the study of the present course of nature may provide, and therefore that we are not authorized, in the infancy of our science, to recur to extraordinary agents."

Lyell convinced his readers that he was taking the only reasonable course, continuing in the Huttonian tradition: Assume that the agents of change that we can observe today have always been the most important ones. Assume that the rates of change have always been rather gradual. Do not recur to extraordinary agents of change. This is the doctrine that has come to be known as uniformitarianism, and it is considered Lyell's great contribution to the geosciences. The publication of *Principles of Geology* all but banished the catastrophist view from geology. The case for gradualism seemed so strong, and a reliance on a very incomplete geologic record so reasonable, that most geologists and students of the history of life embraced the uniformitarian doctrine of Lyell. *Principles of Geology* went through 12 editions, the last in 1875, so generations of geologists were trained in its content.

The general thinking in Lyell's time seems to have been that the answers to the major questions of geology could be found by induction. In other words, when enough facts had been collected, a theory of Earth would emerge. But most of the time, science doesn't work that way. Commonly, someone sees a pattern long before enough data become available to really test the idea. In fact, the theory may specify the kind of data that must be collected.

Geology, when I entered it in the late 1960s, was just experiencing a revolution. The sleepy and dogmatic science had been floundering for years, with no underlying theory to explain Earth's complex features. Was Earth contracting? Was it expanding? Were continents fixed in place or wandering around the globe, or did land bridges once connect the continents, only to subside beneath the waves?

In 1912, Alfred Wegener saw the fit of the continents and the similarity of the rocks and fossils across the southern continents long before the geophysical data from the ocean floor provided the defining evidence of continental drift. This was also true with regard to the role of celestial mechanics in the climatic cycles of the ice ages, where calculations provided the causes and chronology well before convincing evidence was discovered in the geologic record. So science is not like the game show *Jeopardy*, where you win by having the most information at your fingertips, but more like *Wheel of Fortune*, where the first person to solve the puzzle with limited data wins.

But what if Lyell's underlying a priori assumption is actually incorrect? What if the extraordinary changes that we seem to see in the geologic record are really that—extraordinary changes caused by extraordinary events? This thought undermines the classic stories of the early years of our science that we are all told on the first day of Geology 101. Here it was Lyell, a giant of geology, who heroically took on the biblical catastrophists and their perhaps unwitting supporters, such as Cuvier and his colleagues. Now we can see that it was Lyell who advanced a largely theoretical picture of a divinely ordained order of nature, despite the evidence for sudden changes in the geologic record. Lyell argued, wrongly, that rates of known geologic processes have not varied. Violent earthquakes, cataclysmic

volcanic eruptions, flash floods, and tsunamis are all inscribed in the geologic record. Uniformity of process does not imply uniformity of rates.

The most unacceptable kind of extraordinary events is made clear in a
letter written in 1830, in which Lyell maintains that significant changes of
climate can occur "without help from a comet, or any astronomical change."
This jibe was aimed at the musings of scholars such as William Whiston
(1667–1752), the Lucasian Professor at Cambridge, who attributed events
in the geologic record to collisions with comets. In 1696, Whiston published
A New Theory of the Earth, in which he proposed a cosmogony where our
planet originated when a comet was transformed into an ideal world, with a
circular orbit, without tilt or rotation. Later, God sent another comet toward
Earth, and its collision changed the planet's orbit and started it spinning.
This great impact supposedly cracked Earth's crust, releasing the waters of
the Flood, while the vapors of the comet's tail condensed into torrential
rain. Comets were thus transformed from harbingers of local calamities
into natural causes of global cataclysms. Even Edmond Halley and Isaac
Newton surmised that historical, sacred, and geologic periods were punctuated by cometary catastrophes.

Lyell saw that such ideas threatened to lead geology into areas of wild
speculation. Instead, he followed in the footsteps of Hutton, the first to
understand the meaning of "deep time"—the realization that long ages
were required to explain the production of great geologic changes by the
slow and gradual forces of erosion, uplift, and deformation of rocks. It's
clear, however, that Hutton and Lyell did not fully understand the meaning of the deep time they championed. If they had, they could never have
portrayed the history of Earth as dependent primarily on gradual
processes.

Events of many kinds tend to follow a particular inverse relationship
between frequency and magnitude. Small-magnitude events tend to happen much more frequently than large-magnitude events. For example,
small earthquakes are common, larger earthquakes happen less frequently,
and the largest earthquakes are by far the most infrequent, yet these are
the times when geologic changes can most readily be seen. The same holds
true for volcanism (small eruptions are much more common than large
eruptions), for flood events, for landslides, for storms, and for meteorite

impacts. The reasons are variable, but in general an inverse relationship exists between the magnitude and the frequency of events. In the case of meteorite impacts, the individual meteorites are generated by collisions between asteroids in the asteroid belt, which tend to produce lots of small fragments and fewer big ones. In the case of volcanic eruptions, the largest events—flood basalt eruptions—develop during times when huge plumes of hot fluid rock from Earth's interior episodically impinge on Earth's crust.

This means that any study of Earth that includes the notion of deep time must take into account the fact that the greatest-magnitude events should happen very infrequently; in fact, there could be waiting times of millions of years between the largest events. This being the case, the true meaning of deep time is this: even though we expect extremely large events only very infrequently, the long geologic time scale virtually guarantees that such events will happen from time to time, and these energetic events could well be the dominant factors in creating the geologic record.

Studying Earth history with the uniformitarian maxim "the present is the key to the past" ignores the very fact of deep time: insofar as the major events will occur at very long intervals compared with the brief period of our own observations, we will quite likely never see them happen. This is one of the great insights in the geological sciences. We must study the geologic record in order to determine the magnitude of past events that produced that record. In the case of rare meteorite or asteroid impacts, for example, if we determine the population and size distribution of Earth-crossing asteroids and comets, and the record of impact craters on Earth and nearby planets, then we can estimate the waiting times between impacts of various sizes and learn something about the size of the largest possible impact catastrophe.

The religious bent of the early geologists usually is not mentioned when we recognize Dana, Hutton, and Lyell as among the greats of geology. But there is an important theme in geology that runs back to their early work and may be part of the reason why geologists are reluctant to appeal to catastrophic events, even when the geologic evidence points that way. It is a theme that we often overlook or that is present only in an unconscious form. This is the fact that study of Earth is the study of our own planet, so it is difficult to maintain an objective approach when examining its history.

Earth is our home. To Hutton and Lyell, the founders of geology, it was a place created by a beneficent deity to be the orderly, generally peaceful residence of humankind. Why would the planet be forced to undergo periodic catastrophes in which many kinds of life disappear? This would betray the central tenet of God's gift of Earth as the abode for humans. Geologic change was surely not "the work of accident, or effect of an occasional transaction" such as encounters with comets and other extraterrestrial bodies, or volcanic cataclysms of global import.

Geology, it turns out, still retains some aspects of natural theology. Geologists feel close to nature while also standing in awe of the great geologic monuments and the vast expanse of geologic time. I believe that, for many geologists, our science still has deep spiritual and philosophical roots. Are geologists to give up those comfortable feelings of oneness with a peaceful planet, to face an Earth history full of cataclysms and seemingly random destructive events? For most geologists, contemplating the mostly benign effects of slow, grain-by-grain erosion and deposition in altering the landscape, the rhythms of rising and falling seas, and mountains built gradually by uplift and then disappearing slowly by erosion over great periods of time gives subtle intellectual reassurance that the world, if not made for humans, is at least compatible with our survival. Even extinction came as a result of a fair game of competition among species, where the better-adapted organisms survived, and life got better in the process. The biologist David Starr Sloane summed it up beautifully:

> The Universe is with us. It is our Universe and we are a part of it and have no alternative save to accept it as reverently as may be. The positive side of religion is the feeling of being at home in God's world. Whatever our conception of God the attitude remains. God's world to us is no alien land. It is our home and has been the home of our ancestors for aeons immeasurable.

However, new findings in geology suggest that the spiritual feeling of being at home in the universe may be in need of revision.

3

• • • •

The Alvarez Hypothesis

The fact of extinction, much less than the fact of mass extinction,
has been peculiarly difficult for people to accept.

KENNETH HSÜ, *THE GREAT DYING*

In 1973, Nobel Prize–winning chemist Harold Urey published a brief paper in the British scientific journal *Nature* titled "Cometary Collisions and Geological Periods," in which he hypothesized that mass extinctions of life were caused by asteroid or comet impacts. Urey was not the first person to propose such an explanation. If one searches the geological literature, one can find earlier speculations, such as the paper published in 1956 by the invertebrate paleontologist M. W. de Laubenfels in which he proposed that the heat generated by a large meteor impact was involved in the dinosaurs' demise. De Laubenfels made some simple calculations about the heat released from an impacting object, while Urey's paper included back-of-the-envelope calculations of the enormous energy released when a 10-kilometer-diameter (6-mile) object crashes into Earth. Urey suggested that "very great variation in climatic conditions covering the entire Earth should occur and very violent physical effects should occur over a substantial portion of Earth's surface."

Urey proposed that times of impact might be matched to times of extinction and faunal change. Furthermore, Urey's idea was testable. He suggested that geologists look for tektites—glassy objects known to be produced by impacts—at geologic boundaries marked by extinctions, including the end of the Cretaceous period, 66 million years ago, when the dinosaurs disappeared. Even so, at the time, the geological community paid little attention to such a catastrophist idea coming from a chemist. It was outside their uniformitarian paradigm. But, as later developments have shown, one should not be too quick to reject an idea proposed by a Nobel laureate.

Only seven years later, the first major blow to Lyellian doctrine was struck, with the discovery of hard physical evidence that a large comet or asteroid hit Earth 66 million years ago. The initial evidence for this cataclysmic event came from intensive study of a thin band of clay that coincides closely with the last fossils of the dinosaurs and of many other forms of life on land and in the sea. In the late 1970s, Walter Alvarez, a geologist at the University of California, Berkeley, was studying beds of pink marine limestone, called the Scaglia Rossa, now exposed in the Italian Apennine Mountains, in a steep gorge near the lovely medieval/Renaissance hill town of Gubbio. Alvarez came to Italy with Bill Lowrie, a geomagnetism specialist, to investigate the series of reversals in Earth's magnetic field recorded during the deposition of the Gubbio rocks over a period of geologic time stretching tens of millions of years, from the Late Cretaceous to the early Paleogene periods.

Alvarez and Lowrie's idea was to match the patterns of reversals recorded in the Gubbio limestones with the stripes of normally and reversed magnetized ocean crust that were discovered along the mid-ocean ridges. This would enable a cross-check between Earth's magnetism, recorded by the rocks as the ocean floor is extruded along the ridge, and the magnetism acquired by fossil-bearing sediments at Gubbio during the process of their deposition. Such comparison would improve the dating of both sequences. The recording of the reversals in the sedimentary layers is 1,000 times slower (with deposition rates of 1 centimeter [0.4 inch] or less per 1,000 years) than in the magnetic stripes along mid-ocean ridges (with spreading rates of centimeters per year), so that meters of sediment represent millions of years of deposition. Alvarez and Lowrie were particularly interested in potential breaks in the geologic record, which would adversely affect their correlation.

While examining the Cretaceous and overlying Paleogene rock layers, Alvarez noted something peculiar. Most types of Cretaceous marine plankton in the fossil record at Gubbio disappeared in one thin layer of clay (figure 3.1). This level, devoid of fossils, was observed in the 1960s by Italian paleontologist Isabella Premoli Silva of the University of Milan. This

FIGURE 3.1 A close-up of the Cretaceous/Paleogene boundary at Gubbio, Italy. The head of the hammer rests on the uppermost Cretaceous limestone. The thin boundary clay, which has been heavily sampled, is in the recess behind it.

was the boundary between the Cretaceous period of geologic time and the overlying Paleogene period—the so-called Cretaceous/Paleogene boundary, dated at 66 million years ago. In the field, the boundary can be found with just a hand lens, as the numerous millimeter-size planktonic foraminifera (single-celled protists that secrete small-chambered calcareous shells) in the Upper Cretaceous rocks are seen to disappear at a knife edge, right at the boundary clay layer. Alvarez realized that the Cretaceous/Paleogene layer was an important marker in geologic time, where some 75 percent of marine life and the greater part of terrestrial life, including the dinosaurs, perished—a major mass-extinction event.

Above the millimeters-thick bed of barren clay, newer layers of limestone contained a few survivors of the extinction event and new species of ocean plankton that evolved in the wake of the mass extinction. This looked like a potential break in the rock record, and Alvarez wanted to know how long it had taken for the clay layer to accumulate on the seabed. So he took up the problem with his father, Nobel Prize–winning particle physicist Luis Alvarez. Luis was an experimental physicist with a passion for unraveling scientific puzzles. He had worked on the atomic bomb at Los Alamos in the 1940s, observing the first atomic bomb test in Nevada and the bombing of Hiroshima from a chase plane. He even used cosmic rays to search for secret chambers in the Great Pyramid in Egypt.

After some thought, the Alvarezes decided that a good way to estimate the time in which the clay bed had been deposited was to measure the amount of extraterrestrial material in the layer. Small amounts of extraterrestrial debris are continuously raining down on Earth's surface as a result of the burn-up of micrometeorites in the upper atmosphere. This is a very slow process, however, and if the clay layer contained an inordinate amount of this debris, the most straightforward conclusion would be that the thin clay layer was deposited very slowly, over a very long period of time, accumulating more meteoritic dust year by year. If only a little extraterrestrial matter was found in the clay, then it must have been the result of rapid deposition, over a brief period of time.

A strong indicator of extraterrestrial origin would be some element that is normally very rare in the rocks of Earth's crust but that is found in some greater abundance in meteorites. The Alvarezes decided that the rare element iridium would be the ideal tracer of meteoritic fallout. Most of the iridium in Earth resides in its metallic core; it exists in crustal rocks in almost undetectable concentrations. Iridium has a strong affinity for metallic iron, and thus when most of the dense iron sank to the center of the molten early Earth, it carried the iridium with it. By contrast, most meteorites are pristine bits of cosmic matter; they still contain their original amounts of iron and iron-loving elements such as iridium. Hence, meteoritic debris should be rich in iridium, and the boundary clay, if deposited slowly, should be somewhat enriched in the rare metal.

Alvarez, father and son, called on the talents of two nuclear chemists at the Lawrence Berkeley National Laboratory, Frank Asaro and Helen Michel, who were adept in the technique of using neutron activation analysis to measure extremely small amounts of various trace elements. In this type of analysis, the samples are first bombarded by neutrons in an atomic reactor. Atoms of iridium absorb some of the neutrons and then emit gamma rays at a known rate and energy. The nature of these emissions enables the detection of amazingly small amounts of iridium, down to the parts per trillion range, meaning detection of one atom of iridium among a trillion other atoms in the sample.

The results of the chemical analyses surprised the Berkeley group. The Gubbio boundary clay contained more than 30 times as much iridium as even very slow deposition would allow—up to a few parts per billion iridium over a background of only about 30 parts per trillion iridium (figure 3.2). Furthermore, the ratio of the various trace elements in the boundary clay was the same as in common forms of meteorites. The only possible explanation seemed to be that the excess of iridium had been deposited by some event that brought an inordinate amount of meteoritic material to the surface of Earth from outer space.

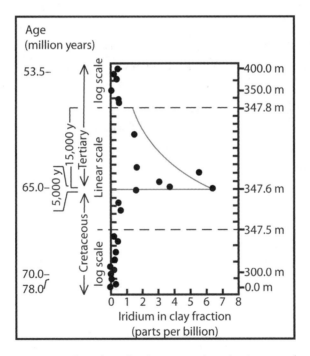

FIGURE 3.2 The iridium abundance anomaly at the Cretaceous/Paleogene boundary at Gubbio. The anomaly occurs over a thickness of about 20 centimeters (8 inches). (Redrawn from L. W. Alvarez et al., Extraterrestrial cause of Cretaceous/Tertiary extinction: Experimental results and theoretical interpretation, *Science* 208 [1980]: 1095–1108)

The Alvarez team also obtained samples of the boundary clay layer from Denmark. Here the clay was somewhat thicker, but it still seemed to represent a short period of time and also coincided with the mass extinction. The analytical results of the Denmark clay were even more surprising: iridium enrichment of the layer was more than 160 times that of the normal sediments above and below the mass-extinction boundary. Samples of the boundary clay obtained by Alvarez from the other side of the world, in New Zealand, also showed iridium enrichment, proving that the layer was a worldwide phenomenon. The Alvarez group came to the dramatic conclusion that Earth must have been hit by a large asteroid or comet, which spread iridium-rich dust around the globe.

This discovery galvanized the geological community. The Berkeley scientists, along with a number of other research groups, began to analyze samples of the boundary clay layer from all over the globe, eventually at more than 350 sites. In most cases, when the composition of the boundary clay was determined, it showed anomalous increased iridium. Many of these sections are in deep-sea sediments like the Gubbio layer, but some are in sediments that were deposited in shallow continental shelf waters, and a number of the sections that have yielded iridium anomalies are found in beds of sedimentary rocks that were once sands and muds on the swampy plains inhabited by the dinosaurs (figure 3.3). The global geochemical anomaly provided solid evidence that a large amount of extraterrestrial matter had fallen from the skies all over Earth just at the time of the global mass extinction.

I remember the first time I saw the boundary layer at Gubbio. I was struck by the counterintuitive conclusion that such an enormous event, which had such a great effect on the history of life, could be represented in the geologic record by a seemingly insignificant clay layer. How many other similar impact layers lay hidden in the rocks? When would the next Earth-shattering impact occur? Since then, I have traveled to more than 25 end-Cretaceous rock sections in Europe, the Gulf Coast of Mexico, the Caribbean, and the western United States, and I am always astounded by the cosmic significance of the thin, seemingly inconsequential boundary clay layer.

The tremendous fiery blast that accompanied the impact would have vaporized the iridium-rich impacting body, excavated a huge crater, and

FIGURE 3.3 Alan Hildebrand sampling the Cretaceous/Paleogene boundary layer from terrestrial deposits in the Raton Basin of New Mexico.

created a widespread dust cloud. A cloud so massive that it would have completely blotted out the sun. The iridium-enriched dust would have been globally distributed, eventually falling out of the atmosphere to form the base of the worldwide clay layer. The almost total darkness and cold stemming from the cloud of dust in the atmosphere, and soot from wildfires set by hot debris reentering the atmosphere and lighting up the skies, should have led to a total loss of photosynthesis and a break in the planetary food chain, killing off a majority of existing species of animals and plants on land and in the sea. From the calculation of the amount of iridium in the clay worldwide, and knowledge of the average composition of meteorites, the Alvarez team estimated that the diameter of the world-changing asteroid or comet was about 10 kilometers (6 miles), similar to the estimate considered earlier by Urey.

Most of Earth's atmosphere is less than 10 kilometers from the surface, so an invading body traveling at 20 to 60 kilometers (12 to 37 miles) per second would have penetrated the atmosphere in a fraction of a second; there would be no warning of the impact. The dinosaurs would not have seen it coming. The extraterrestrial body smashed into Earth with such force that it created extremely high pressures and temperatures. The impactor and target rock were turned to vapor, melted, and pulverized, and the ultrahigh pressures created characteristic shock-induced minerals, such as shocked quartz (figure 3.4). Shocked quartz grains show telltale sets of microscopic, closely spaced, crossing parallel lines at specific crystalline angles and can be produced only by the intense pressures generated by a hypervelocity impact; volcanoes will not do the trick. The energy released in the impact spewed the tiny grains of quartz all over Earth. In 1983, Bruce Bohor and his colleagues at the U.S. Geological Survey in Denver discovered shocked quartz grains in the Cretaceous/Paleogene boundary clay in the western United States, adding direct, unequivocal evidence for an impact.

Also seen in the boundary layer are tiny pellets of various compositions, which in many cases were originally made of glass (many are altered but still have glassy cores). First observed in samples of the boundary clay in Spain by Jan Smit, a geologist at the Free University in Amsterdam, these are microtektites, tiny versions of the glassy tektites that Urey said might be found at geologic boundaries (figure 3.5). They average about

FIGURE 3.4 A shocked quartz grain from the Cretaceous/Paleogene boundary. The multiple crisscrossing planes of fracture are diagnostic of shock. The grain is about 0.1 millimeter (0.004 inch) across.

FIGURE 3.5 Microtektites (up to a few millimeters in diameter) from the Cretaceous/Paleogene boundary in Haiti. (Photo by David Kring)

0.1 millimeter (0.004 inch) in size, but the largest may be several millimeters in diameter (large enough to be considered tektites). These small beads were originally droplets of melted rock, splashed out of a large impact crater and distributed widely, and their commonly spherical and teardrop shapes attest to their aerodynamic transport. Some tiny spheres in the boundary clay appear to have formed directly through rapid condensation and coagulation of rock vaporized in the impact; these nonglassy beads are called microspherules.

Smit also had a curious role to play in the discovery of the iridium anomaly. In the late 1970s, he was working on the Cretaceous/Paleogene boundary in Spain. He, too, was struck by the sudden disappearance of the tiny shells of ocean plankton at a thin clay layer. Smit and a colleague made an elemental analysis of the Spanish boundary layer, but he was not initially made aware of the iridium abundance—it seemed much too high to be correct. By the time Smit realized that there was an iridium anomaly (he was recovering from an illness), the Alvarez group had announced their discovery. Smit published his paper in the British journal *Nature*, and it actually appeared before the Alvarezes' paper came out in *Science*, but Smit acknowledged the Alvarez group's priority. Walter Alvarez has said many times that he considers Smit to be a codiscoverer of the iridium anomaly.

The rebirth of catastrophism may have been in the air. In 1979, on the eve of the Alvarezes' discovery, a prescient account of an impact/extinction event appeared in the book *Earth Shock* by geologists Basil Booth of Imperial College and Frank Fitch of the University of London. They proposed that "water vapor and dust generated by a large impact would rise into the upper atmosphere and would influence the world's weather for a considerable period." According to their book, "It is not impossible that the sudden demise of certain species in the past—for instance the dinosaurs—may be attributable to such a cosmic catastrophe." Impact catastrophe's time had come.

In a sidelight, in 1982, clay mineralogist Bob Reynolds at Dartmouth and I obtained from Smit samples of the Cretaceous/Paleogene boundary clay from several sites. We wanted to look at the clay to determine which minerals were present. We hoped that some exotic minerals might occur in the layer that would give a clue to its origin. We studied the fine fraction of our samples (the fine-grained clay minerals) by X-ray diffraction analysis but did not look at the coarser fractions, where the shocked quartz grains and microtektites would have been. So we missed finding the evidence of impact.

Shocked quartz and microtektites are clear indications of impact, and their discovery in the Cretaceous/Paleogene boundary layer proved that the layer was produced by fallout from a great impact somewhere on Earth. If there were any lingering doubts, in 1989, impact expert John McHone and a team from Arizona State University discovered stishovite, a high-density phase of quartz found only at impact craters, providing even more solid evidence that the clay layer represented impact crater fallout. Further studies of the boundary layer found microdiamonds, abundant soot, and non-terrestrial amino acids, all related to the impact and its aftereffects.

By 1990, the major missing piece of the puzzle remained the impact crater itself—an enormous crater that had evaded detection. Impact simulations and collision experiments suggested that the crater produced is typically about 20 times the diameter of the impactor itself. According to these calculations, an approximately 180- to 200-kilometer-diameter (112- to 124-mile) crater dating from 66 million years ago should be present somewhere on Earth. You would think that a crater that size could be easily found, but over time the evidence fades. Craters are eroded or covered by more recent sediments. Of course, ocean impacts would be difficult to detect, and the crater could have been excavated on a part of the ocean floor that has since been subducted and therefore destroyed.

At the time that the Alvarez group reported the discovery of the impact layer, in 1980, no such crater was known anywhere on the planet. One clue to the crater's location was the presence of the shocked quartz in the impact layer. The mineral quartz is associated with continental rocks like granite, so a continental impact seemed most likely. The largest shocked grains in the Cretaceous/Paleogene boundary layer at the time were found in geologic sections in the western United States. These sites showed the presence of a spike of fungi and fern spores at the boundary, marking the early recovery of pioneering flora after the cataclysm. The western U.S. boundary clay also was somewhat thicker than the European sections, suggesting that the impact had occurred somewhere in the Western Hemisphere. Studies from around the Gulf Coast of the southern United States and Mexico, and from some Caribbean islands, showed coarse deposits of sand and disturbed beds exactly at the Cretaceous/Paleogene boundary (figure 3.6). Jody Bourgeois of the University of Washington first studied these beds in coastal Texas. An expert in the sedimentary record of tsunamis, Bourgeois found evidence in the type of sediments and their internal structures that the Gulf Coast beds had been deposited by giant waves more than 100 meters (328 feet) high.

At Arroyo el Mimbral in eastern Mexico (see figure 3.6), at a site that was submerged below the Gulf at the time, the Cretaceous/Paleogene boundary is marked by 2-meter-thick (7-foot) bed of coarse sand full of fossilized wood. Beneath the sand bed, a thick layer of altered microtektites marks the time of the impact (figure 3.7). Just above the massive sand bed, layers of ripple-marked sand suggest rapidly moving currents of water that reversed direction from time to time, and, above those layers, one finds the iridium anomaly and shocked quartz grains, marking the later fallout of material launched high above Earth. The coarse sand layers are interpreted as the deposit of a tsunami that washed up along the coasts and drew

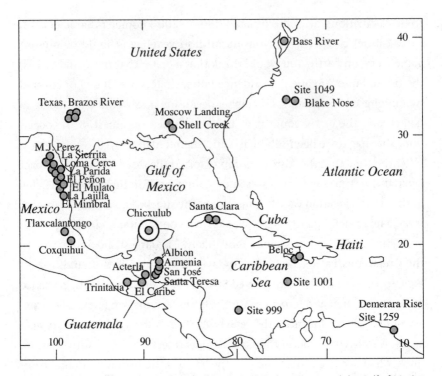

FIGURE 3.6 The locations of Cretaceous/Paleogene impact deposits around the Gulf of Mexico and the Caribbean Sea. The location of the Chicxulub crater is indicated. Tsunami-related deposits are widespread in the area.

sediment from coastal beaches and mangrove swamps down into the deep basin of the Gulf. The ripple-marked sands represent deposits apparently formed as the tsunami waves sloshed back and forth within the Gulf of Mexico.

Great submarine landslides around the Caribbean also occurred in the wake of the impact. This is understandable, because such a giant impact would generate at least a magnitude 10 earthquake, with considerably more energy than the largest historic quakes. Seismic effects would have been felt across a wide area. In western Cuba, for example, I have seen that the sedimentary record of the boundary event is a single massive bed some 40 meters (131 feet) thick, with coarse debris at the bottom and fine-grained sediments at the top, which was deposited en masse when a shallow-water carbonate platform and reef collapsed into deeper water, triggered no doubt by a huge impact-induced earthquake.

FIGURE 3.7 The Cretaceous/Paleogene boundary at Arroyo el Mimbral in eastern Mexico. The prominent 2-meter-thick (7-foot) sandstone layer represents a tsunami deposit.

During the late 1980s, Canadian geologist Alan Hildebrand, then a brash young graduate student at the University of Arizona, also explored the Texas Gulf Coast tsunami deposits. Hildebrand was making a concerted search for the crater in the Gulf of Mexico/Caribbean region. The geologic trail led him to Haiti, where Cretaceous and Paleogene strata had been studied by Haitian geologist Florentin Maurrasse of Florida International University. Florentin is a keen-eyed expert on ancient fossils and sedimentary rocks; he and I were students together at Columbia in the early 1970s. In the late 1970s, Maurrasse discovered in Haitian outcrops a 50-centimeter-thick (20-inch) deposit of coarse sand, which he called a volcanic turbidite— a layer formed by a fast-moving undersea slurry of sediment down the slopes of a submarine volcano. Maurrasse found this turbidite layer exactly at the Cretaceous/Paleogene boundary, as determined by the microfossils in the sediments.

Hildebrand was intrigued by the apparent thickness and composition of the Haitian layer. Maurrasse had even suggested that the layer contained microtektites, similar to those he had once studied in Upper Eocene sediments from the Caribbean. Anyone putting together Urey's paper on impacts, published in 1973, and the presence of tektites at geologic boundaries in Maurrasse's work, in the late 1970s, could have deduced that there was an impact at the Cretaceous/Paleogene boundary and scooped the Alvarez team—but no one did. Lyell's laws still prevented geologists from seriously considering extraterrestrial impact as a viable mechanism for geologic change.

Hildebrand immediately flew to Haiti with David Kring (from the Lunar and Planetary Laboratory at the University of Arizona) and sampled the thick Haitian boundary layer. They found that it was made mostly of small pellets, some of which contained cores of glass. Recognizable shapes (spheres, teardrops, and dumbbells) agreed with the aerodynamic forms exhibited by tektites and microtektites (see figure 3.5). The boundary layer also contained large shocked quartz grains. Hildebrand realized that such a thick layer of impact debris in Haiti meant that the impact structure must be nearby in the Caribbean region.

I visited the Haitian Cretaceous/Paleogene boundary with Maurrasse somewhat later, in 1996. The best outcrop, near the town of Beloc, occurs on a slope so steep that we had to use ropes to reach the spot halfway down the embankment. When I got there, I noticed that the outcrop was crawling with black widow spiders, so we approached our sampling gingerly, while barefoot Haitian kids ran up and down the hillside, curious about what we were doing there.

After he visited Haiti in 1990, Hildebrand's luck continued. A reporter told him about a riveting talk at a meeting of the Society of Exploration Geophysicists in Los Angeles, back in 1981. Petroleum geologists Glen Penfield and Antonio Camargo and colleagues at the Mexican oil company Pemex reported a huge subsurface basin on Mexico's Yucatán Peninsula (figure 3.8). They even suggested that it might be the "smoking gun" for the end-Cretaceous impact. The gaping crater lay well hidden, buried beneath thick overlying sediments. Through drilling into the crater for samples and studying the results of geophysical measurements, such as a determination of Earth's gravity field over the basin, the petroleum geologists were able to map out the true dimensions of the approximately 180-kilometer-diameter (112-mile) circular, multiringed structure (figure 3.9). A report of their

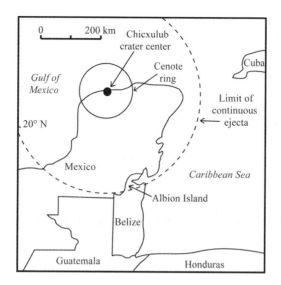

FIGURE 3.8 The location of the Chicxulub impact structure on the Yucatán Peninsula. The continuous ejecta are largely debris-flow deposits. The location of Albion Island in Belize is also shown.

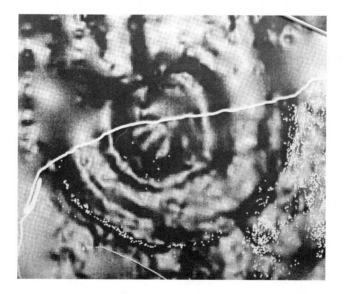

FIGURE 3.9 A gravity anomaly map of the Chicxulub impact structure on the northern coast of the Yucatán Peninsula. The white line is the northern Yucatán coastline and the white dots are cenotes (small sinkhole lakes) that follow the crater's outer rim. (From A. R. Hildebrand et al., Size and structure of the Chicxulub crater revealed by horizontal gravity gradients and cenotes, *Nature* 376 [2002]: 415–17)

findings had even been published in the March 1982 issue of the popular magazine *Sky and Telescope* (which I often read), but the proposed mega-crater was somehow overlooked by the entire impact community.

In 1990, Hildebrand managed to obtain some of the precious well cut-tings from a deep layer within the basin, which the Pemex drillers described as andesite, a volcanic rock. One look at the samples under a microscope revealed the telltale shocked quartz and remnants of melted glass. Hilde-brand knew that the huge basin must be the searched-for Cretaceous/Paleo-gene impact structure. The geologists named it Chicxulub (which I'm told means "tail of the devil") after a town in the middle of the buried struc-ture. I would have chosen the nearby town of Progresso; it is much easier to pronounce, and progress of a monumental order had been made for the impact hypothesis.

At the same time as Hildebrand's discovery, Jet Propulsion Lab geolo-gists Adriana Ocampo and Kevin Pope contributed airborne evidence. Using satellite images of the Yucatán, they found that cenotes, numerous small sinkhole lakes, on the peninsula traced out a great arc that actually followed the circumference of the Chicxulub structure (see figure 3.9). The lakes tracked the edge of the crater, where deep arcuate fractures allowed under-ground waters to reach the surface and dissolve some of the overlying limestone.

Of course, the discovery of the crater generated new disagreements. Some researchers argued that the age of the crater differed from the age of the boundary clay and hence that they were not related. But Carl Swisher at Berkeley and his colleagues, through careful radiometric dating in a single lab, soon showed quite convincingly that Chicxulub melted rock and Hai-tian tektites were exactly the same age. Tom Krogh of the Royal Ontario Museum in Canada and his colleagues designed a beautiful study based on their discovery of tiny shocked grains of the mineral zircon in the bound-ary clay. These grains, dated by uranium-lead radiometric methods, recorded two events—one around 540 million years ago for the origin of the zircons, which matched the age of the target rock granite basement at Chicxulub, and an overprinted age of 66 million years ago, corresponding to the impact shock event.

Critics also proposed that the glass might be volcanic in origin, but a group of researchers led by well-known volcanologist Haraldur Sigurdsson at the University of Rhode Island found that the glass from the impact site and the glassy centers of the microtektites in Haiti had the same unusual calcium- and sulfur-rich composition. They are different from any known volcanic glass but are in accord with the impact target in the Yucatán, with thick beds of limestone (calcium carbonate) and anhydrite (calcium sulfate), which would have been incorporated into the melted rock.

With the discovery of Chicxulub crater and its timing, the impact hypothesis for the end-Cretaceous extinction gained wide acceptance in the geologic community. Thirty years of work on the Cretaceous/Paleogene impact/extinction problem was summarized in a paper published in 2010 in *Science*. Forty-one coauthors with expertise in various fields pertinent to the question of impacts and mass extinction all supported the idea that an impact had caused the extinctions. They also concluded that the crater that marked the world-changing impact was indeed the 180-kilometer-diameter (112- mile) Chicxulub structure.

Theories, however well substantiated, always have their detractors. Some dinosaur experts continued to maintain that the dinosaurs died off gradually over a long period of time. They argued that the numbers of dinosaur fossils could be interpreted to show a progressive decline in the last few million years of the Cretaceous period. The resolution to this disagreement most likely lies in the fact that the geologic record is capricious and our sampling is imperfect, and hence the preservation and discovery of dinosaur fossils is a chance occurrence. Statistically, one can show that finding the last dinosaur skeletons exactly at the impact layer would be very unlikely.

Thus sudden events can appear gradual. For example, at the time of the discovery of the iridium layer, the last large dinosaur bones found in Upper Cretaceous beds in the western United States were about 3 meters (10 feet) below the anomaly. These Cretaceous sediments were deposited at variable rates, but geologists estimated that the 3 meters of sediment would have taken more than 10,000 years to accumulate. On the face of it, it appeared that the last dinosaurs disappeared many millennia prior to the impact event. Dinosaur fossils, however, especially large articulated skeletons, are only rarely preserved in the geologic record. Therefore, considering the spotty record of fossilization and sampling, the dinosaurs could have lived right up to the boundary anomaly.

To check on that idea, in 1991, paleontologists Peter Sheehan and David Fastovsky from the Milwaukee Public Museum, with colleagues, mounted a careful study of the Cretaceous/Paleogene boundary beds in Montana and North Dakota. The study made use of a cadre of volunteer geologists to sample and identify every scrap of bone in the uppermost Cretaceous layers. They found no evidence of a gradual dinosaur die-off. What's more, at several localities in the western United States, geologists have found dinosaur footprints—clear evidence that dinosaurs were present—within a few tens of centimeters (about 1 foot) below the boundary layer, but never above it. The viewpoint of some paleontologists who rejected the impact/extinction hypothesis is revealed by Norman MacLeod of the Natural History Museum in London, who observes that "the secret of a productive life in science is to have a chronic insoluble problem, and to keep working on it." For many scientists, however, the goal is to solve the problem and to move on to its implications.

In the case of microfossils, like the small calcareous plankton (foraminifera and coccoliths with shells of calcium carbonate) that make up the Gubbio limestones, sampling and smearing of the fossil record are less of a problem, because each sample contains myriad tiny fossils. One would expect, however, that some rare species might not be sampled in the last beds below the iridium anomaly and that other species might extend into the Paleogene because of reworking of fossils from older beds. We know, for example, that burrowing organisms can mix sediments from above and below the boundary clay. Despite this mixing, the calcareous plankton show a major sudden extinction event coincident with the iridium layer in many sections around the world.

At least one group of geologists, led by micropaleontologist Gerta Keller at Princeton, proposes that all the extinctions did not occur at the same time and that a number of marine species survived the Cretaceous/Paleogene event. Their ideas are based on their own studies of the boundary clay and deep drill cores from the Yucatán crater. Keller does not believe that the

Chicxulub crater and the extinction are related. She has steadfastly maintained that the Chicxulub impact took place 300,000 years before the Cretaceous/Paleogene boundary as defined by microfossils and the iridium anomaly. Recently, however, Paul Renne of Berkeley and his colleagues showed, through precise age determinations, that the ages of the boundary clay in the western United States (dated by volcanic ash layers close to the boundary) and of the impact glass of the Haitian boundary layer date to within less than 37,000 years of each other. This is within the error bars of the two dates, making them, essentially, exactly the same age.

The weight of opinion with regard to the connection between the impact and the mass extinction now seems overwhelming. But Keller's group is sticking to its guns, also claiming that the tsunami deposits along the Gulf Coast (see figure 3.6) were actually the result of great changes in sea level, which seem to be recorded nowhere else. The bottom line may be that any geologic sections relatively near the impact (such as the Gulf Coast and Caribbean deposits) are likely to have been disturbed by tsunami and seismic activity, so that their beds give a distorted and inaccurate picture of the events at the boundary.

The Cretaceous/Paleogene boundary layer is meters thick along the Gulf Coast of Mexico and the United States, a few centimeters thick in the western United States, and only a few millimeters thick in Italy, falling off with distance from the crater. But what would the impact debris look like closer to the impact site in the Yucatán? In 1996, I joined a group from the Planetary Society (led by Ocampo and Pope) doing fieldwork on the Cretaceous/Paleogene boundary at Albion Island in Belize. This exposure of boundary rocks is one of the closest to the Chicxulub crater (that we know of), hundreds of kilometers from the crater's outer ring. The deposits that we found there are unlike any others previously seen at the boundary.

The Planetary Society's expeditions to Belize provided a unique look at the by-products of one of the most dramatic processes set in motion by the massive impact—ballistic debris flows. During the impact, fragments of the target rocks were ejected upward and outward from the crater at high velocity. The debris traveled in arcing paths, eventually falling back to Earth. Imagine a curtain of such debris, sweeping rapidly outward from the impact site, violently striking the ground, tearing up the surface rocks, and moving along as a ground-hugging flow of mixed ejecta and local rock. Such a storm of rock would have leveled everything in its path, burying the countryside under tens of meters of debris, until finally the flows ran out of steam, hundreds of kilometers from the crater.

The Belize localities are the closest that scientists can get to the Chicxulub crater to study the ejected material, without having to drill through a thick cover of younger rocks. At Albion Island, the Cretaceous/Paleogene boundary rocks are exposed in a large quarry, where the uppermost Cretaceous Barton Creek Dolomite is being mined. The Cretaceous/Paleogene boundary deposits overlie the fractured Barton Creek beds (figure 3.10). At the base of the boundary deposits we found an impact layer about 1 meter (3 feet) thick, full of green altered spherules and containing shocked quartz. This layer is immediately overlain by a spectacularly chaotic mixture of materials of all sizes—the debris-flow deposits—from a matrix of fine clay to large limestone and dolomite boulders meters across. Geologists have traced these coarse boundary deposits across Belize, and similar chaotically

FIGURE 3.10 The Cretaceous/Paleogene boundary at Albion Island, Belize, about 300 kilometers (185 miles) from the Chicxulub crater. The fractured latest Cretaceous Barton Creek Dolomite is overlain by a meter of darker sediment containing spherules and shocked quartz (at the heads of Kevin Pope on the left and David King on the right), which in turn is overlain by the impact breccia, made up of debris-flow deposits of Chicxulub ejecta.

mixed sediments have been found at a number of other Central American sites. At some sites in Belize, the boulder-filled debris-flow deposits are overlain by lowest Paleogene marine sediments, fixing their age as latest Cretaceous.

Such impact breccias as they are called ("breccia" being a geologic term for coarse-grained sediments mostly made up of angular fragments) are found in and around several known terrestrial impact craters. For example, the Bunte Breccia, surrounding the 15 million-year-old, 26-kilometer-diameter (16-mile) Ries crater in southern Germany, contains a chaotic mixture of the multicolored target rocks of the impact. Other impact-related sediments almost certainly lie hidden in the geologic record, interpreted as debris-flow deposits of various origins. We will come back to this question in a later chapter.

All in all, the Alvarez hypothesis has been extremely fruitful in explaining the deposits seen at the Cretaceous/Paleogene boundary all around the globe. These sediments form a reference in the search for impact evidence at other times of extinction in the geologic past.

4

• • • •

Mass Extinctions

Paleontologists have sometimes gone beyond this descriptive phase of the subject and have attempted to formulate the "causes," "laws" and "principles."

T. H. MORGAN, *A CRITIQUE OF THE THEORY OF EVOLUTION*

Earth is apparently a very dangerous place. The geologic record is littered with the remains of extinct animals and plants. In fact, it is estimated that more than 99 percent of all the species that have ever lived on Earth are gone forever. More than 10 million modern species have been identified, and many more wait to be discovered and classified. But the fossil record contains at least hundreds of millions of extinct species. Indeed, for most species, existence is relatively brief—only a few million years on average. After that, they disappear without issue or die out after giving rise to some new evolutionary branch. What's more, species don't always die out one at a time, as isolated cases; they often disappear during mass-extinction events in which large numbers of existing species go extinct suddenly.

Why should organisms become extinct? In the view of Charles Darwin in *On the Origin of Species* (1859), extinctions occur when two species are in competition for a similar ecologic niche, and only one species wins the competition. (Note that this is species selection, not natural selection per se.) Darwin used the analogy of a series of wedges driven into a log. These are the species that are adapted to their particular niches. For Darwin, competition is like the driving of a new wedge into the log, which invariably causes one of the other wedges to pop out. In Darwin's view, extinctions are always taking place as part of the ongoing "struggle for existence."

What about mass extinctions? According to Charles Lyell and to Darwin, who followed him, they don't exist. Lyell insisted, in *Principles of*

Geology, that if large numbers of species disappear in going from one geologic formation to the next, it just means that the intermediate beds, which would have shown that the species became extinct gradually and one at a time, were missing, the result of erosion of rock layers or of nondeposition in the first place. Darwin bought into Lyell's conclusion that mass extinctions were an illusion, created by an incomplete geologic record. Darwin resolved, in *Origin of Species*, that "the old notion of all the inhabitants of Earth having been swept away at successive periods by catastrophes is generally given up. . . . Species and groups of species gradually disappear." Lyell even surmised that the full length of time that was missing between the Cretaceous and Paleogene beds, based on the dissimilarity of the fossils, must be as long as the entire time since the Cretaceous/Paleogene boundary, which we now know is 66 million years. Clearly, this much time cannot be missing from the boundary sections. In fact, studies of the magnetism of the boundary rocks by Dennis Kent at Lamont-Doherty Earth Observatory and Rutgers University revealed that any missing time at the boundary must be less than 10,000 years, at the limits of geologic temporal acuity.

The case for impact causing the end-Cretaceous extinction is strong, but the mass extinction at the end of the Cretaceous is not the only one in the geologic record, and not even the most severe. Starting in the late 1970s and on into the early 1980s, while the impact hypothesis for the end-Cretaceous event was being hotly debated, Jack Sepkoski, a paleontologist at the University of Chicago, was working calmly on a valuable database of the diversity of life. Sepkoski earned a doctorate under Stephen Jay Gould at Harvard University. In 1978, he went to Chicago, and he was also a research associate at the Field Museum of Natural History. Unfortunately, in 1999 he died at the top of his field, succumbing to heart failure at the age of 50.

A brilliant and hardworking paleontologist, Sepkoski's primary field area was the library. He is perhaps best known for his global databases of the time ranges of marine animal families and genera, data sets that continue to motivate paleobiological research. In his initial compilation, he tabulated the stratigraphic intervals of apparent origination and extinction of more than 3,500 animal families in the marine record of the past 600 million years. Sepkoski initially chose the family taxonomic level because more detailed records, at the genus or species levels, were thought to be less reliable. For the basic time intervals, he used the well-marked geologic stages (on average about 5 or 6 million years long), which in most cases were naturally defined by a turnover in the fossil fauna.

Sepkoski spent years assembling his database. He started his compilation using standard sources (primarily *Fossil Record* and *Treatise on Invertebrate Paleontology*), but he soon graduated to using the primary literature buried in hundreds of volumes of individual fossil studies—and this was before the Internet. The paleontological literature is vast and is presented in many languages, and the publications go back more than 200 years. In 1982, Sepkoski published *A Compendium of Fossil Marine Animal Families*. The collection of data immediately proved useful, and among Sepkoski's first discoveries was support for the apparent exponential increase in diversity of life at the beginning of the Cambrian period—the so-called Cambrian explosion, 542 million years ago.

The new compilation of data provided the basis for new analyses of marine diversity and of the evolutionary history of particular groups of marine organisms. Sepkoski's *Compendium* showed that, since the start of the Cambrian period, the record of fossil diversity has been punctuated five or six times by major mass extinctions, some even worse than the end-Cretaceous event, in terms of species loss (figure 4.1). These extinction events were followed by recovery periods during which new organisms apparently refilled emptied ecological niches. In some cases, these rebound periods seem to have taken millions of years. The extinction data also showed a number of less severe extinction peaks.

It is difficult to see how Darwin's ideas of extinction caused by competition could explain mass extinctions. Some have maintained that ecosystems are so finely balanced that the extinction of a few key species could cause a complete breakdown, a chaotic event. Most biologists, however, do not think that species are so deeply intertwined, especially between land and sea. Over the years, many proposed hypotheses have considered mass extinctions to be caused by physical changes in the environment. For example, some scientists have suggested that changes in global climate, sea level, or volcanic activity could have led to climatic changes, causing

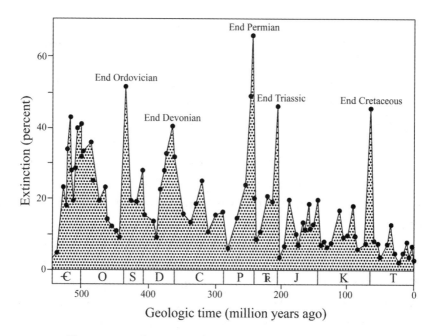

FIGURE 4.1 The percentage of extinctions of marine genera over the past 540 million years. The graph shows the five or six major mass extinctions and about 20 lower peaks. These genera extinction values can be readily converted to species extinction rates. The geologic periods, from oldest to youngest, are Cambrian, Ordovician, Silurian, Devonian, Carboniferous, Permian, Triassic, Jurassic, Cretaceous, and Tertiary. The five major mass extinctions are the peaks that occur at the end of the Ordovician (444 million years ago), the Late Devonian (372 million years ago), the end of the Permian (252 million years ago), the end of the Triassic (201 million years ago), and the end of the Cretaceous (66 million years ago). Another potential major extinction peak is found in the Late Permian period (260 million years ago).

less-well-adapted species to die out, with new ones eventually taking their place. Others believe that movements of the continents may have allowed competitive mixing of species that had been separated by oceans, or that a reduction in continental shelf area after continental collisions, or at times of low sea level, could lead to increased competition and species extinction. But these are mostly gradual mechanisms, and the mass extinctions are seen to happen suddenly. Changes in sea level, climate, and volcanism occur quite often in the geologic record, but it seems that abrupt extinction events require more violent processes.

Sepkoski eventually produced a more detailed compilation, at the genus level in taxonomy and at the finer substage level in geologic time, that he made available to the geologic community. Jack was always very open about sharing his data. He sent me his latest version of the extinction database at the genus level in September 1991 and told me to "have fun with the data." Other compilations of the comings and goings in the fossil record come from Mike Benton's book *The Fossil Record 2* (1993), which contains results for 7,186 families of marine and nonmarine organisms, and more recently from the Paleobiology Database, which came online in 2001. That data set records the occurrences of genera and species from within particular fossil collections. Today, the database includes more than 18,000 genera and is continually updated. A study conducted in 1995, using *Fossil Record 2*, and another in 2008, using the database, agreed quite well with Sepkoski's original data.

The genera and family extinction data can be extrapolated to species-level extinction using a method called rarefaction. This method was developed by Sepkoski's close collaborator, paleontologist David Raup at the University of Chicago, and is based on the fact that the extinction of a genus requires the extinction of all species in that genus. So, for example, a 45 percent extinction of genera translates into a 75 percent extinction of species (see figure 4.1). Raup specialized in quantitative paleontology. He was apparently more comfortable at a computer screen than out in the field collecting fossils. (In 1980, Raup reviewed the Alvarezes' paper for *Science* and rejected it, but he made a number of suggestions for improving the manuscript.) Using the extinction data, Raup showed that the five or six major extinction events were not just an extension of the lesser-magnitude events; they were exceptional events well outside the envelope of the smaller events.

The most severe mass extinction occurred at the end of the Permian period of geologic time (about 252 million years ago). Sepkoski's data, with Raup's rarefaction technique, shows that about 96 percent of extant species with shells became extinct in the seas, and similar numbers of extinctions were estimated for land-based life. Other sources show that reptiles, land plants, and insects were especially hard hit; the forest trees died out suddenly, leading to a devastated landscape. In addition to the end-Permian mass extinction, the other high peaks of extinctions occurred at the end of the Ordovician period (444 million years ago), during the Late Devonian period (including the Frasnian/Famennian boundary [372 million years ago]), at the end of the Triassic period (201 million years ago), and, of course, at the end of the Cretaceous period (66 million years ago) (see figure 4.1).

These are the "big five" mass extinctions recognized by paleontologists for some time. There are some other relatively high peaks of extinction within the Cambrian period (542 to 485 million years ago), but they are best interpreted as artifacts of the relatively sparse fossil data from that interval. Another high extinction peak occurs in the Late Permian (Guadalupian stage), at 260 million years ago, but that peak may be partly an artifact of sampling and a smearing backward of species terminations from the massive end-Permian extinction, which occurred only 8 million years later. Sepkoski's data show elevated extinction levels in the Late Devonian period, lasting for about 30 million years. Sepkoski also was able to pick out other times of lesser extinction events that were still significantly above the background level of extinctions, commonly marking the boundaries of geologic periods and epochs (see figure 4.1).

With the strong evidence for an impact as the cause of the end-Cretaceous extinction, it would seem natural for scientists to ask whether these other extinctions also had been caused by extraterrestrial impacts. Instead, some geologists rejected the idea out of hand and claimed that other mass extinctions—such as the very severe end-Permian event—were long-drawn-out affairs attributed to purely terrestrial causes. Despite considerable evidence to the contrary, some geologists have continued to argue that mass extinctions of life result from myriad terrestrial causes, such as species competition, changes in climate and sea level, explosive volcanism, and even chaotic collapse of ecosystems. But major and rapid changes in climate, sea level, and volcanism during the latest ice age, in the past 2 or 3 million years, did not cause significant extinctions on land or in the oceans.

Some geologists have even proposed that an unusual concatenation of various geologic events conspired to produce the major mass extinctions. This is the "Murder on the Orient Express" model, named by paleontologist Douglas Erwin at the Smithsonian Institution, after Agatha Christie's famous multiple-murderer mystery. The best explanation for the mass extinctions, its proclaimers say, involves random, purely terrestrial processes working in concert. This would mean that processes usually "down in the noise" conspire to create an extinction signal.

The plethora of geological explanations reminds me of the situation prior to the acceptance of plate tectonics, when geologists spouted all sorts of hypotheses to avoid accepting continental drift as the most likely process. I see this rejection of impact as a viable candidate for the cause of extinctions to be part of the bias that can be traced back to Lyell's limiting of the geologic record to slow, earthbound processes. But the simplest explanation is not that normal, ongoing geologic processes conspire in some way to produce extraordinary outcomes. The simplest explanation is that extraordinary results are produced by extraordinary causes, whether terrestrial or extraterrestrial. This is just the application of Occam's razor.

The distribution of the extinction events in time is also noteworthy. The idea that extinctions of life may be periodic goes back to a paper published in 1952 by catastrophist paleontologist Norman Newell of the American Museum of Natural History. Newell suggested that periodic changes in sea level affected the distribution and diversity of life in shallow seas and on the continental shelves.

Sepkoski did not assemble his fossil compilation in order to look for cycles, but when he shared the new data set with Raup, they noticed that the extinctions over the past 250 million years seem to have had a periodic distribution with an approximately 30 million–year cycle. In an earlier study, done in 1977, noted geologist Al Fischer and his student Mike Arthur, then at Princeton, suggested that biological crises and associated changes in ocean circulation and climate came in a roughly 32 million–year cycle. Sepkoski's paleontological data, however, were much more complete and accurately dated, and in order to test for the reality of an approximately 30 million–year extinction periodicity, he teamed up with Raup for a careful statistical study of the mass-extinction record.

In the early spring of 1983, Raup and Sepkoski did the number crunching of the extinction data, revealing the presence of a significant periodicity of 26 million years in extinctions over the past 250 million years. They first presented their results at a conference called "Patterns of Change in Earth Evolution," held in Berlin in May 1983, but Sepkoski's presentation was so low key that the press did not pick up on it. In August of that year, however, Sepkoski gave a talk at the "Dynamics of Extinction" symposium in Flagstaff, Arizona, where several reporters noted the revolutionary nature of the study, and the story appeared in *Science, Science News,* and the *Los Angeles Times* in September.

I read the early news reports of Raup and Sepkoski's findings and, realizing their implications, wrote to Raup and asked for a copy of the unpublished article. I received it a few days later. So I had a copy of the radical but seemingly solid research well before publication. If their findings were true—if extinctions were periodic and the end-Cretaceous event was caused by an impact—then the other extinctions in the series may also have been caused by impacts, and large-body impacts may also be periodic, possibly driven by some extraterrestrial cycle. This was exciting stuff.

Raup submitted his extinctions manuscript to *Proceedings of the National Academy of Sciences,* where, because he was a member of the academy, external review could be dispensed with and publication could be very rapid. Little did he know that the wheels of research were already spinning; his unpublished paper had set in motion a wave among geologists and astrophysicists, which we will discuss in chapter 11. The study marked the start of a debate focused on whether mass extinctions were periodic. In one of the first critical papers published, in *Science,* Stephen Stigler, a well-known statistician at the University of Chicago, and his coauthor, Melissa Wagner, challenged the data and time series analyses. They pointed out that the geologic timescale used by Raup and Sepkoski to date the extinctions contained a bias toward a period of about 26 million years. However, Stigler and Wagner, not being geologists, did not realize that the geologic timescale is not just a series of dates but was created by analyzing fossils from strata all over the world and noting times when changes took place. Later, these times were dated by radiometric methods and interpolation of ages. Thus the timescale has the record of extinctions built into it.

My colleague Richard Stothers and I subsequently analyzed the extinction record, using several competing timescales, and found that, in all cases, the extinctions showed a 26 million–year periodicity. Raup and Sepkoski looked at only the past 250 million years, where the dating of these events was considered most reliable. More recently, with improved age determinations, I found evidence that the approximately 26 to 27 million–year extinction cycle goes back 540 million years (figure 4.2), and this was verified by physicist Adrian Melott of the University of Kansas

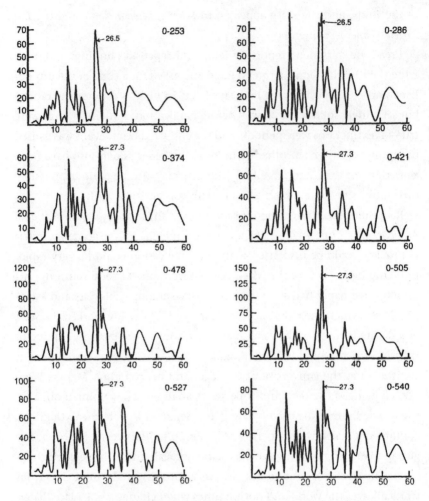

FIGURE 4.2 Spectral analyses of the record of extinction events going back 540 million years. The horizontal axes represent the period in millions of years, and the vertical axes represent spectral power. Spectral power remains high at a period of 26.5 to 27.3 million years.

and paleontologist Richard Bambach of the Smithsonian Institution in 2013, using the latest revised geologic timescale back to the Cambrian period.

Science excels in scrutiny, and over the years, Raup and Sepkoski's results have been revisited by a number of researchers, utilizing different methods of spectral analysis and various subsets of the extinction data. Several researchers examined the data and concluded that, even if the extinctions were periodic, the available data were too meager to detect it. Others, such as Steven Stanley of Johns Hopkins University, proposed that the time distribution of mass extinctions merely reflected "the existence of postcrisis recovery intervals that space out the mass extinctions, preventing them from following in close succession," creating a pseudo periodicity. It is doubtful, however, that such occurrences would produce the statistically significant 26 million–year periodicity seen in the extinction record. Raup eventually concluded that about half of the subsequent studies supported the periodicity argument and half found either the statistical methods or the available data wanting.

More recently, Coryn Bailer-Jones of the Max Planck Institute for Astronomy in Heidelberg performed sophisticated spectral analysis of the extinction data and did not find a significant periodicity. But Bailer-Jones admits that it is difficult to detect a periodicity in data composed of mixed periodic and nonperiodic events, as is most likely the case for the extinctions. Several years ago, Doug Erwin at the Smithsonian Institution observed that "the periodic signal continues to shine through the turmoil, battered but resilient."

There the record stood, until 2015, when I teamed up with my former student Ken Caldeira, now at Stanford, utilizing a new method of time series analysis. Ken was the perfect graduate student; he could take any idea and run with it. Now he advises Bill Gates on issues of climate change and energy policy. The study, published in the *Monthly Notices of the Royal Astronomical Society* in 2015, again found a significant periodicity of 26 million years in extinction events over the past 250 million years. I am sure, however, that we will continue to draw fire from critics of periodicity. It seems that the jury is still out on the question of cycles in the extinction data.

But if the extinctions are periodic, it leads to some very interesting questions. We have seen the evidence for impact at the time of the end-Cretaceous extinction, and the record of other extinctions. Could these other mass-extinction events have been caused by large-body impacts? Is our dangerous world caught in a regular cycle of destruction? Before taking on those questions, it is worthwhile to consider some of the predicted environmental effects of large-body impacts on Earth.

5

• • • •

Kill Curves and Strangelove Oceans

A physicist can react instantaneously when you give him some evidence that destroys a theory that he previously had believed. But that is not true of all branches of science, as I am finding out.

LUIS ALVAREZ, "EXPERIMENTAL EVIDENCE THAT AN ASTEROID IMPACT LED TO THE EXTINCTION OF MANY SPECIES 65 MILLION YEARS AGO"

Earth from space is a beautiful blue marble. But our planet sits within a swarm of Earth-crossing asteroids and comets, and collision of these bodies with Earth represents an ongoing natural astronomical and geologic process (figure 5.1). We need only look at the face of the moon to see the effects of impacts; the larger Earth provides an even bigger target and suffered even more impacts. Astronomical observations of Earth-crossing comets and asteroids of various sizes can be used to estimate the expected average times between collisions of large or small objects with Earth. Geologists must look up as well as down to unlock the story of our changing planet.

Estimating collision rates can also be accomplished independently by counting craters of various sizes on Earth, the moon, and other planets. The numbers obtained by both methods, the asteroid survey and the crater counts, agree very well. Many more small bodies than large bodies enter the count (as one would expect from objects created by innumerable collisions in space), so that small impacts are common and large impacts are relatively rare. Gene Shoemaker of the U.S. Geological Survey Astrogeology Science Center in Flagstaff, Arizona, one of the world's experts on impacts and impact craters, used these counts to create a chart of the estimated intervals of time between impacts of various sizes (figure 5.2). Shoemaker, an inspiring pioneer in impact studies, first proved that Meteor Crater in Arizona is indeed the scar of an impact. The crater had generated

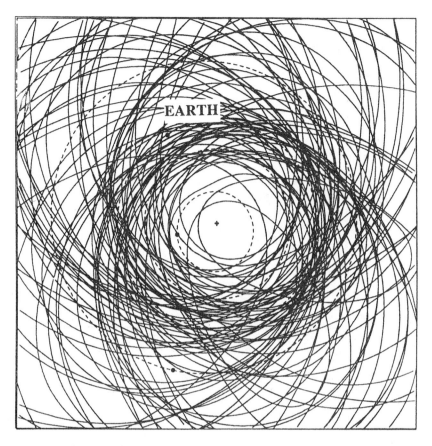

FIGURE 5.1 The orbits of the 100 largest Earth-crossing asteroids and the orbit of Earth. Our planet sits within a "shooting gallery."

speculation for more than 100 years. Was it a volcanic crater or the result of an extraterrestrial event? In the 1960s, Gene provided the clinching evidence for impact—high-pressure phases of mineral quartz. Tragically, he was killed in an automobile accident in Australia in 1997 while tracking down new craters in the outback. His ashes now rest on the moon.

In figure 5.2, we can clearly see that small impact events are predicted to happen quite frequently, in geologic terms. And yet even small impacts can have dramatic effects. The collision of the 50-meter-wide (164-foot) object that caused the Tunguska explosion in the sky above Siberia in 1908 created a blast so fierce that it scorched and flattened a large area of forest

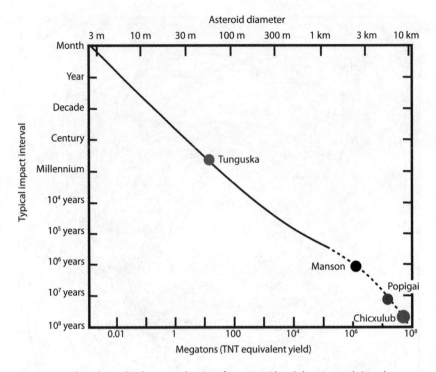

FIGURE 5.2 The relationship between the size of an asteroid and the expected times between impacts of various sizes. Manson is a 40-kilometer-diameter (25-mile) crater, Popigai is a 100-kilometer-diameter (62-mile) crater, and Chicxulub is a 180-kilometer-diameter (112-mile) crater.

(figure 5.3). The Tunguska bolide exploded high in the atmosphere with energy equivalent to a few megatons (million tons) of TNT, the energy equivalent of a nuclear device. Such explosions could take out a city. Luckily, most Tunguska-like explosions occur over the oceans and in remote areas and so have not been observed. The Chelyabinsk fireball, which exploded in midair over Russia in 2013, shattering thousands of windows, was an even smaller object, only a few meters across. What if ancient legends reporting "fire and destruction from the sky" are descriptions of the impacts of small asteroid and comet fragments?

On the large end of the size spectrum, the asteroid/comet data and cratering records on the planets predict that Earth will be hit by a 10-kilometer-diameter (6-mile) object every 100 million years on average, and over that time period we would experience a number of impacts of objects a few

FIGURE 5.3 The site of the Tunguska blast of 1908 in Siberia. This destruction of forest was caused by the explosion of a small asteroid or comet, 50 meters (164 feet) in diameter, in the upper atmosphere.

kilometers across (see figure 5.2). The rise in the curve for large impacts represents a contribution from the relatively large comets. A collision with such objects has predictable far-reaching effects that would constitute regional or global environmental disasters. On top of that, surges of asteroids, resulting from the breakup of large main-belt asteroids, should have occurred from time to time. As we shall see, brief increases in the comet flux above the background level, lasting a few million years and resulting from gravitational perturbations of the distant Oort cloud comets, could also occur.

Truly enormous amounts of energy are released when large asteroids and comets, traveling at cosmic velocities of 20 to 60 kilometers (12 to 37 miles) per second, collide with Earth. The kinetic energy of an impactor is proportional to its mass times its velocity squared. In such a collision, a 10-kilometer-diameter (6-mile) asteroid or comet, traveling at tens of kilometers per second, would release energy equal to the incredible explosion of about 100 million megatons of TNT. This is the equivalent of about 1 billion atomic bombs of the kind that were dropped on Japan at the end of

World War II. No wonder many life-forms vanished at the end of the Cretaceous.

In 1986, Shoemaker estimated that comets, which are generally larger than asteroids, might account for more than 50 percent of the largest impact craters. So, according to his chart, the past 542 million years should have seen five or six impacts of 10-kilometer-diameter (6-mile) objects, which seem capable of causing major global extinctions. Such major extinctions would consist of 75 percent to 95 percent species loss. Furthermore, during that same period Earth would have been bombarded by approximately 25 impacts of 5-kilometer-diameter (3-mile) objects, or about one every 22 million years. These would carry 10 percent of the energy of a 10-kilometer object and probably would be capable of causing less-severe extinctions—about 30 to 50 percent species loss. Shoemaker's data speak powerfully of a planet under siege.

By comparison, Jack Sepkoski's record of extinctions of marine species over the same period shows five or six major peaks of extinction and around 19 minor peaks, also about one every 22 million years (see figure 4.1). When two numbers agree—one from astronomy, for the expected frequency of large impacts, and the other from paleontology, for the frequency and severity of extinction events—it is time to take a closer look at the data. On the face of it, these statistics suggest that it is just possible that all of the extinction peaks identified in Sepkoski's data could be related to large-body impacts. This is a radical idea.

David Raup was one of the first to suspect this, and he decided to combine the extinction data with information on the time between terrestrial impact craters of various sizes to construct a potential relationship between mass extinction and impact cratering, a so-called kill curve (figure 5.4). He proposed that the relationship would follow an S-shaped, or sigmoidal, curve form, where small impacts have little effect while large impact events would cause mass extinctions. Eventually, a saturation level is reached for very large impacts—the remaining few percent of resistant living species would be difficult to kill off, even with a massive impact event. Geologist and author Peter Ward of the University of Washington pointed out that this quantitative kill curve concept represents one of the most powerful ideas to emerge from the entire extinction debate.

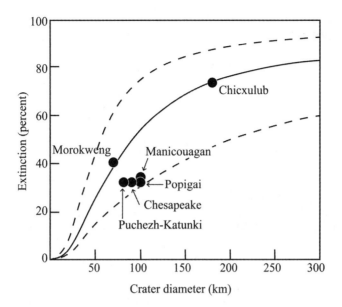

FIGURE 5.4 David Raup's kill curve (with dashed estimated error bars) for marine species. Plotted are the six largest impact craters of the past 260 million years, all of which are coeval with extinction events: Chicxulub (66 million years [Mexico]); Morokweng (145 million years [South Africa]); Manicouagan (214 million years [Canada]); Popigai (36 million years [Russia]); Chesapeake (36 million years [United States]); and Puchezh-Katunki (168 million years [Russia]).

Although it is clear that the severity of an impact-induced mass extinction would probably depend on a number of variables (including ambient climate conditions, susceptibility of fauna and flora, and the target site of impact), the size and energy of the impactor are likely to be among the most important factors. When I compared Raup's theoretical curve with known large, well-dated impact craters of the past 250 million years (measuring 70 kilometers [43 miles] in diameter or more), the observed points in the figure agreed with the predicted kill curve within the envelope of error permitted by the geologic data (see figure 5.4). As I have recently shown in a study with Ken Caldeira, all six of these large impact craters seem to be represented by known impact-derived layers in the geologic record and correlate with extinction events (table 5.1), a very unlikely situation, statistically.

The shape of the impact/species kill curve predicts that, for individual craters smaller than about 60 kilometers (37 miles) in diameter, there will

TABLE 5.1 The Six Largest Impact Craters (Greater Than 70 Kilometers [43 Miles] in Diameter) of the Past 260 Million Years, with Corresponding Stratigraphic Evidence of Impacts and Extinction Events

Crater	Diameter (kilometers/miles)	Age of Crater (millions of years)	Impact Ejecta	Extinction Event
Chesapeake	90/56	35.3 ± 0.1	Iridium anomaly, microtektites, shocked quartz	Late Eocene
Popigai	100/62	35.7 ± 0.2	Iridium anomaly, microspherules, shocked quartz	Late Eocene
Chicxulub	180/112	66.04 ± 0.05	Iridium anomaly, microspherules, microtektites, shocked quartz	End-Cretaceous
Morokweng (Mjølnir*)	$\geq 70/43$ (40/25)	145 ± 0.8 (142.6 ± 2.6)	Iridium anomaly, microspherules, shocked quartz	End-Jurassic
Puchezh-Katunki	80/50	167.3 ± 3	Iridium anomaly, microspherules	Bajocian
Manicouagan	100/62	214 ± 1	Iridium anomaly, microspherules, shocked quartz	Middle Norian

*The ejecta layer is from the roughly coeval Mjølnir impact crater.

be no associated extinction pulse that stands above the 20 to 25 percent background level of global species extinction. And, indeed, there are well-dated craters of this size range that apparently do not correlate with global extinction peaks in Sepkoski's data. They may, however, coincide with lesser geologic boundaries based on minor or regional faunal turnover. Furthermore, the points in the figure also could be interpreted as indicating a possible step in the kill curve at a crater size of about 70 to 100 kilometers (43 to 62 miles), suggesting some kind of threshold effect.

Raup had earlier provided some evidence for an extinction threshold based on computer simulations of asteroids of different sizes hitting Earth, producing "lethal radii" within which a large percentage of species are killed off. Using a definition of a mass extinction as elimination of at least 30 percent of marine species, Raup found that a lethal radius of about

10,000 kilometers (6,200 miles), or about half of the surface of Earth, is required. Anything less would be lost in the ongoing background of global extinctions but might cause enough faunal turnover to be represented as a lesser geologic boundary.

"Nuclear winter" climate modeler Brian Toon and his colleagues at NASA's Ames Research Center used computer simulations to study the effects of large impacts on the atmosphere. Their work allowed a quantitative estimate of the threshold impact energy—around 70 million megatons of TNT—required to cause global dust clouds, equivalent to the effect of an impact of a comet or an asteroid several kilometers in diameter. Toon's work suggests that the dust cloud from such an impact could almost completely cut off solar radiation for months, reducing global atmospheric transmission below the minimum required for photosynthesis. A similar threshold exists for continental-scale wildfires, set off by hot ejecta reentering the atmosphere, broiling Earth's surface. Climate simulations by Toon and others predict that land surface temperatures would drop catastrophically, by about 30°C (54°F) in less than a week, under such heavy dust cloud conditions—the so-called impact winter.

The potential connection between climate change and mass extinctions is a much-debated subject, with some arguing that the extinctions were related to a cooling of the climate, while others have implicated warming in mass-extinction scenarios. One must remember, however, that the effects of impacts on Earth's climate could be short-lived, creating a few months to years of impact winter, or could be long term, lasting thousands or even millions of years, especially if positive feedbacks from ice sheet growth or involvement with the oceans come into play. Estimates of connections between impacts and climate have been approached in two ways: theoretical and modeling studies that attempt to estimate the kinds and magnitudes of climatic and geologic changes that impacts of various sizes might induce, and study of proxy environmental and climatic indicators at times of documented or suspected large-body impacts and extinction events, like the Cretaceous/Paleogene boundary.

For example, the Alvarez group initially estimated that the duration of the global dust cloud from the Chicxulub impact, and its effects, would have been several years, based on atmospheric opacity measurements after historical volcanic eruptions like that of Krakatau in 1883. However, as my work on volcanoes and their effects on climate back in the 1980s showed, such relatively long-lived volcanic cooling events are the result not of dust but of tiny droplets of volcanically produced sulfuric acid. The somewhat larger dust grains mostly settle out within a few months, as would impact dust, whereas the stratospheric sulfuric acid aerosols produced by Krakatau, and the global cooling they caused, apparently persisted for a few years.

Was there an extended cooling after the sharp impact winter conditions predicted for the boundary event? The Yucatán impact site provided a mostly limestone target (made of calcium carbonate), but about 25 percent of the sediments is in the form of calcium sulfate (the mineral anhydrite). This type of sediment is called an evaporite—the anhydrite salts were deposited at times when seawater in shallow seas in the Yucatán was evaporated in a warm, dry climate. So, in addition to the dust, a great deal of the vaporized sulfate would have been added to the upper atmosphere. There, the sulfate could be converted to tiny droplets of sulfuric acid by reaction with atmospheric water. It is therefore possible that a fine global aerosol cloud formed (and perhaps reached much higher altitudes than material from volcanic explosions), which could have stayed in the atmosphere for years, reducing incoming solar radiation and cooling the climate. Geologists Wenbo Wang and Tom Ahrens of Caltech estimated that the sulfate release from the Chicxulub impact could have created enough aerosols to produce a large global cooling of up to 10°C (18°F), lasting for several years.

In the case of an ocean impact, water would probably represent a significant fraction of the material launched to high altitudes. In the case of a 10-kilometer-diameter (6-mile) asteroid or comet hitting the ocean, the sudden expansion of a huge steam bubble would drive a mixed plume of vaporized asteroid or comet, solid debris, and water vapor into the upper atmosphere. Water vapor is a strong greenhouse gas and could have led to significant warming. Conversely, condensation of water vapor and its subsequent freezing could produce ice particles in the upper atmosphere, and ice clouds could have reflected a considerable amount of solar radiation back to space, further cooling the climate. Positive feedback from highly reflective snow and ice ground cover also could have led to extended cold conditions.

With such a scenario in mind, geochemist Frank Kyte of UCLA proposed in 1985 that the small asteroid (about 0.5 kilometer [0.3 mile] in diameter) that produced the Eltanin oceanic impact in the South Pacific about 2.3 million years ago, at an especially climatically sensitive time, may have been enough to trigger the present ice age.

Shock waves produced by the passage of the comet or asteroid through the atmosphere should have led to reactions between atmospheric oxygen and nitrogen, creating nitrogen oxide compounds. The global atmosphere after an impact would have been loaded with nitrate (1,000 times more than during the heaviest air pollution episodes today), which is quickly converted to corrosive nitric acid. The resulting highly acidic rainfall would add to the postimpact destruction. Nitric acid created by the impact, along with sulfuric acid from the anhydrite target rock, would have increased the acidity of soils and ocean surface waters. The net results could have been widespread destruction of plants on land, and perhaps an acidic surface ocean uninhabitable by many calcareous organisms.

The material from the crater, launched on ballistic tracks, some outside the atmosphere, interested cratering expert Jay Melosh, now at Purdue University. He performed calculations of the heat emitted as ejected material reentered the atmosphere and heated up from atmospheric friction. He found that the pulse of heating from incoming ejecta might reach 50 to 100 times the solar radiation (equivalent to setting your oven on broil) for periods of up to several hours. This brief heating event should have been enough to kill exposed animals and ignite standing vegetation, producing widespread wildfires. The soot produced by the flash fires would have added to the opacity of the atmosphere, in the style of a nuclear winter, exacerbating the darkness and the cooling after a large impact. Not surprisingly, large amounts of soot (estimated at 100 billion tons worldwide) were discovered at the Cretaceous/Paleogene boundary in many localities, which supports the burning of a significant fraction of the biomass on the continents.

The Cretaceous/Paleogene boundary is also marked by a major perturbation in the global carbon cycle, with a decrease in oceanic life and productivity and a change in the deposition of carbon in the ocean (figure 5.5). The carbon cycle involves the transfer of carbon between the solid Earth and the ocean–atmosphere system. During normal times, the release of

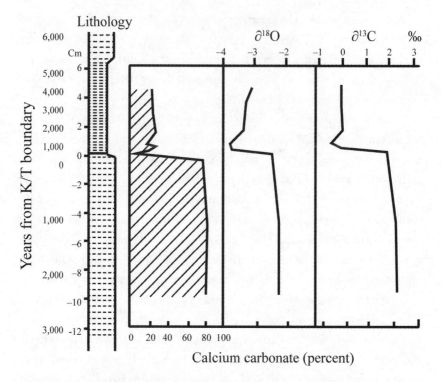

FIGURE 5.5 The percentage of calcium carbonate and the carbon and oxygen isotope records across the Cretaceous/Paleogene boundary in Spain. Note the negative shift in the carbon and oxygen isotopes and the marked reduction of calcium carbonate with the extinction of calcareous plankton at the boundary.

carbon dioxide into the atmosphere comes mainly from volcanism. During the end-Cretaceous event, however, impact into the carbonate platform of the Yucatán could have instantaneously released large amounts of carbon dioxide directly into the atmosphere. John O'Keefe and Tom Ahrens at Caltech estimated that the increased atmospheric carbon dioxide would have created a significant greenhouse effect and would have raised global temperatures by 4°C to 10°C (7°F to 18°F) for 10,000 to 100,000 years after the impact.

Further potential for long-term global warming lies in the fact that marine calcareous phytoplankton are known to release the gas dimethyl sulfide, which oxidizes in the lower atmosphere to form tiny droplets of sulfuric acid. These fine aerosol particles are the precursors of cloud condensation nuclei over the oceans. Cloud condensation nuclei are needed as seeds for cloud development. A decrease in such nuclei would result in a decrease in cloudiness and a reduction in the reflectivity of clouds over the oceans. More sunlight would reach Earth's surface, which would lead to climate warming. This becomes interesting for the end-Cretaceous event, as almost all calcareous plankton disappeared at that time. Once the looming worldwide dust cloud subsided, the skies may have been mercilessly cloud free. This shows how an extinction of plankton in the sea could affect animals on the land.

In 1988, at New York University, climate scientist Tyler Volk and I calculated the changes in cloud reflectivity with variations in cloud condensation nuclei, and the surface temperature changes that might then occur. Given a dimethyl sulfide reduction of 90 percent, we predicted that global temperatures would rise by nearly 10°C (18°F). These results suggest that a drastic reduction in dimethyl sulfide production could have led to a substantially warmer Earth following the mass-extinction event, which may have persisted until calcareous phytoplankton populations recovered, which may have taken up to tens of thousands of years. These high temperatures could, in fact, have been a factor inhibiting the recovery of the marine and terrestrial biospheres.

Several other lines of evidence point to a trauma in the global biosphere at the Cretaceous/Paleogene boundary. At major geologic boundaries, coinciding with extinction pulses, high-resolution stratigraphic techniques make it possible to identify a global horizon of mass killing, marking a time at which a large proportion of the biomass disappears over a very brief interval—less than 10,000 years, or perhaps almost instantaneously. This could lead to a significant anomaly in the ratio of carbon isotopes (carbon-12 and carbon-13) preserved in sediments across the Cretaceous/Paleogene boundary. The data show evidence for an increase in light carbon (carbon-12) in surface oceans and the atmosphere, indicated by analyses of the calcareous shells of surface-dwelling plankton (see figure 5.5). At times of normal ocean productivity, light carbon in ocean surface waters is depleted because abundant photosynthetic plankton preferentially use the light carbon and their organic remains sink to the seafloor, taking the light carbon with them. If surface ocean photosynthesis decreases drastically, then the light carbon is no longer being preferentially taken up by the phytoplankton and it accumulates in the surface waters, where it is incorporated into the shells of calcareous plankton and deposited as limestone. Thus one plausible explanation for a large increase of light carbon in the shells of calcareous plankton is a great reduction in surface phytoplankton productivity.

The light-carbon anomaly at the Cretaceous/Paleogene boundary seems to be a worldwide phenomenon, and it suggests a surface ocean of very low productivity, the so-called Strangelove ocean, named after the film *Dr. Strangelove or, How I Learned to Stop Worrying and Love the Bomb*, about global nuclear holocaust. These dead-ocean conditions may have lasted for a long time. In the South Atlantic, for example, the light-carbon anomaly begins just above the boundary (marked by the iridium-rich layer) and reaches a minimum value about 50,000 years later.

Isotopes of oxygen from marine carbonate sediments have been used in a number of studies as an estimator of ocean temperatures at the time of the end-Cretaceous extinctions. Two isotopes of oxygen, heavy oxygen (oxygen-18) and light oxygen (oxygen-16), are found in the carbonate ions (CO_3^{-2}) dissolved in the ocean. In the 1950s, Harold Urey at the University of Chicago first realized that the ratio of light oxygen to heavy oxygen in the carbonate shells of marine organisms varies with the ambient temperature of their surroundings. So the organisms act as thermometers, recording past climatic changes (see figure 5.5).

In the Late Cretaceous, before the mass extinction, several climatic indicators, such as the lack of polar ice and the presence of palm trees and cold-blooded reptiles in near-polar latitudes, suggest a global climate significantly warmer than today's. Ocean surface waters at that time may have been up to 5°C (9°F) warmer than at present, with warm waters even at subpolar latitudes. Variations in oxygen isotope ratios at the Cretaceous/Paleogene boundary occur in many places in the world ocean (see figure 5.5) and suggest that a sudden warming of the surface oceans of up to 10°C (18°F) took place across the boundary, closely correlated with the maximum variations in carbon isotope ratios. This heating could have been due to a combination of increased atmospheric carbon dioxide and a decrease in cloudiness.

Evidence from fossil plants also shows a significant rise in atmospheric carbon dioxide and warming across the boundary. Using the inverse relationship between atmospheric carbon dioxide and the number of stomata (small pores) on the surface of fossil plant leaves, David Beerling from the University of Sheffield and his colleagues reported evidence for a marked increase in carbon dioxide (up to six times the present atmospheric carbon dioxide level) at the Cretaceous/Paleogene boundary. This much carbon dioxide in the atmosphere would have increased the greenhouse effect and may account for some of the observed global warming. Recent work based on leaf shapes of plant fossils (which vary with temperature) in the western interior of North America also supports a long-term warming of about 10°C (18°F) across the Cretaceous/Paleogene boundary in that area, an increase that seems to have persisted for hundreds of thousands of years after the impact event.

During a mass extinction, a host of species that were well adapted over time to their particular ecological niches suddenly disappears. In the aftermath of these extinction events, opportunistic species suddenly become widespread (in the postimpact ocean at the end of the Cretaceous, for instance, several kinds of unusual plankton proliferated), and some surviving species undergo what is called an adaptive radiation, in which the survivors, under the guidance of natural selection, adapt to fill the many recently vacated niches. This applies to the amazing evolutionary radiation of mammals and birds in the wake of the end-Cretaceous extinctions. Over a relatively brief period of geologic time, the available niches are filled, and once the niches are filled, it becomes difficult for a new species to wedge another species out of its niche. During those more stable times, most new variations in a species would be deleterious and would be weeded out by stabilizing selection, and the stable ecosystem would continue.

This pattern in the fossil record—with sudden extinction, abrupt speciation events, and long periods of stasis—lines up with the evolutionary model called punctuated equilibrium, as proposed by paleobiologists Stephen Jay Gould at Harvard and Niles Eldredge at the American Museum of Natural History in the early 1970s. But Gould and Eldredge were thinking of the punctuated evolution of individual species. The concurrent and sudden disappearance of many species in mass extinctions bespeaks a larger-scale phenomenon that might be called "punctured equilibrium."

Although much has been written about the effects of large impacts, smaller impacts also may have produced significant changes on a regional scale. For example, collisions of 2- to 3-kilometer-diameter (1- to 2-mile) asteroids and comets (with energies in the millions of megaton range), producing impact craters from 40 to 60 kilometers (25 to 37 miles) in diameter, should occur about every million years, or about 500 to 600 such impacts in the past 542 million years. About 30 percent of these impacts (approximately 200) would have hit the continents, which means that, for example, North America, representing about 15 percent of total continental area, would have suffered around 30 such impacts, or about one every 20 million years. Eleven craters in this size range are known in North America. Calculations suggest that these energetic events should have left a significant imprint on regional geologic records. In the case of a million-megaton event, the radius of blast devastation could be hundreds of kilometers from the impact site, and the radius of fire ignition might be as great as 1,000 kilometers (620 miles).

A good example of this kind of impact is the 35-kilometer-diameter (22-mile) Late Cretaceous Manson impact structure in northwestern Iowa (dated at about 74 million years ago). An estimated 2 million–megaton impact of a 2-kilometer-diameter (1-mile) body produced this crater. It is now completely buried but has been extensively drilled. Geologist Glen Izett of the U.S. Geological Survey, an expert on shocked minerals, searched for Manson ejecta in Upper Cretaceous marine rocks to the west of the structure. He discovered shocked quartz in a widespread layer in southeastern South Dakota (the Crow Creek bed), which has been interpreted as a probable impact-tsunami deposit. The equivalent Pine Ridge Sandstone in eastern Wyoming, which I visited in 1999, also looks like a tsunami deposit, with rafts of offshore sediment deposited where large waves washed up along the ancient shoreline of the Cretaceous seaway.

For decades, scientists have debated whether dinosaurs in western North America were declining in diversity prior to the end of the Cretaceous period. The confusion may be the result of a regional extinction caused by the Manson impact only 8 million years earlier. Coincident with the

Manson event, nine genera of dinosaurs disappeared from western North America, followed by the immigration and rapid evolution of other dinosaurs and by major changes in marine reptiles and tiny mammals in the same region. In 2002, I proposed that catastrophic dinosaur bone beds in Montana may mark the far boundary of the Manson impact event. Sites of mass mortality contain herds of thousands of individuals in rocks dated at about 74 million years ago. Were they killed by the blast wave? Or did fires ignite their world, which lay more than 1,000 kilometers (620 miles) from Manson? A Late Cretaceous fossil forest site in northwestern New Mexico, 800 kilometers (500 miles) from the impact site, may also have felt the destruction caused by the Manson impactor. Thus a number of puzzling features of the Late Cretaceous extinctions in western North America may be explained by the regionally devastating effects of the Manson impact event.

While studying impact crater data, I discovered something interesting about their spatial distribution. In the midwestern United States are eight circular geologic structures, ranging from about 3 to 17 kilometers (2 to 10 miles) in diameter, showing evidence of outward-directed radial deformation and intensely brecciated rock. What's more, these circular features are lined up. They lie within a narrow linear swath about 15 kilometers (9 miles) wide along a straight line stretching about 700 kilometers (435 miles) across the United States, from southern Illinois through Missouri to eastern Kansas (figure 5.6). Two of these features, the Decaturville and Crooked Creek structures, are regarded as bona fide impact craters, as they are known to contain shocked quartz and shatter cones—horsetail-like impressions left by shock waves passing through the rocks, which occur only in impacts. The other six structures, which have not been carefully studied with impact in mind, contain uplifted, deformed, and brecciated country rock (with forceful intrusions of breccia into the surrounding rocks), shattered quartz grains, and material that is interpreted as recrystallized melt glass.

Based on their similar geologic characteristics and the presence of diagnostic and/or probable evidence of shock, these structures, once classified as elusive "cryptovolcanic" or "cryptoexplosion" structures (involving internal explosive processes of a kind that has never been observed), are more confidently seen as resulting from hypervelocity impact. No other

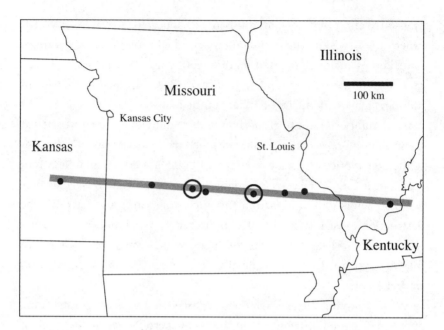

FIGURE 5.6 The linear distribution of eight circular structures in the midwestern United States showing severe deformation. *From left to right*: Rose Dome, Weableau, Decaturville, Hazel Green, Crooked Creek, Furnace Creek, Avon, and Hicks Dome. The Decaturville and Crooked Creek structures (*circled*) are certified impact structures.

similar occurrence of aligned features is known on Earth, and I calculated the probability of a chance alignment of the eight structures to be less than one in a billion. The unusual alignment suggests that the features are coeval and related to a multiple impact event, with a best-constrained age of about 330 to 310 million years ago, a time of minor extinctions. Calculations suggest that the proposed impact crater chain is unlikely to have been formed by an incoming comet or asteroid disrupted by terrestrial or lunar tidal effects, but may have been the result of a string of asteroidal or cometary objects produced by breakup somewhere within the inner solar system, reminiscent of the impacts of fragmented comet Shoemaker-Levy 9 on Jupiter in 1994.

When I proposed the impact origin of these structures, some years ago, I thought it was obvious that they must be related, considering the very unlikely alignment and the fact that two of the structures were already identified with impact. I was surprised that several geologists (one was the reviewer of my paper on the structures) wrote comments again attributing the structures to never-observed cryptoexplosions and ignoring the evidence for impact. The features were originally called the "38th parallel structures" because they are now closely parallel to that latitude, though of course they would have had a quite a different orientation in the Paleozoic, when they formed. Again, I think Lyell's uniformitarian laws are still preventing some scientists from seeing the obvious.

6

• • • •

Catastrophism and Natural Selection

CHARLES DARWIN VERSUS PATRICK MATTHEW

The real core of Darwinism is the theory of natural selection. This
theory is so important for the Darwinian because it permits the
explanation of adaptation, the "design" of the natural theologians,
by natural means.

ERNST MAYR, *TOWARD A NEW PHILOSOPHY OF BIOLOGY*

Charles Darwin (1809–1882; figure 6.1) sits in the pantheon of great scientists for his discovery that evolution takes place by the process of natural
selection. Darwin was a gentleman scientist like Charles Lyell. His family
wealth allowed him the free time to study evolution and the history of life.
Darwin followed his mentor Lyell in considering that all change, both geologic and biological, was gradual. It is a hallmark of Darwin's theory that
evolution is continuous and takes place in small steps, as natural selection
chooses among small variations in the forms of life. Thus microevolution
leads to macroevolution in a continuous manner. Darwin attributed extinctions of individual species primarily to interspecific competition. Darwin,
again following Lyell's lead, interpreted the apparent revolutions in the history of life as illusions caused by gaps in the stratigraphic record. Sedimentary rocks either were not formed during those gaps or were later eroded
away. Those missing sedimentary rock layers, if present, would show that
the extinctions had occurred gradually and were spread over a considerable period of time.

It is of interest to examine Darwin's views of gradual evolution with
respect to the recent picture of geologic change marked by catastrophes.
Furthermore, natural selection is such a powerful explanatory principle (it
has been called "the best idea anyone ever had") that the paths leading to
its discovery have attracted significant scholarly analysis, and a veritable
industry has developed in tracing the origins of Darwin's thoughts on the

FIGURE 6.1 Charles Darwin (1809–1882), author of *On the Origin of Species*.

idea. The first public announcement of the concept came in papers by Darwin and Alfred Russel Wallace, read in absentia at London's Linnean Society in 1858, followed by the publication of Darwin's monumental *On the Origin of Species* the next year. Darwin's notebooks show that he arrived at the idea in 1838, and he composed an essay on natural selection as early as 1842, but he refrained from publishing until he discovered that Wallace had come to similar conclusions.

Much earlier, however, in 1831, the Scottish horticulturalist Patrick Matthew (1790–1874; figure 6.2) published a remarkable pre-Darwinian formulation of the law of natural selection (he actually used the phrase

FIGURE 6.2 Patrick Matthew (1790–1874), author of *On Naval Timber and Arboriculture*.

"natural process of selection"), largely in an appendix to his book on forestry, *On Naval Timber and Arboriculture*. In his brief exposition of natural selection, Matthew wrote:

> There is a natural law universal in nature, tending to render every reproductive being the best possibly suited to its condition. . . . As the field of existence is limited and pre-occupied, it is only the hardier, more robust, better suited to circumstance individuals, who are able to struggle forward to maturity, these inhabiting only the situations to which they have superior adaptation . . . the weaker, less circumstance suited, being

prematurely destroyed. This principle is in constant action, it regulates the colour, the figure, the capacities, and instincts; those individuals whose colour or covering are best suited to concealment or protection from enemies, . . . whose figure is best accommodated to health, strength, defence, and support . . . those only come forward to maturity from the strict ordeal by which Nature tests their adaptation to her standard of perfection and fitness to continue their kind by reproduction.

Matthew's prescient publication, hidden as it was in the appendix of a little-read specialist volume, did not generate debate on the subject at the time. But, in 1860, after the publication of the *Origin*, Matthew wrote a letter to the journal *Gardeners' Chronicle and Agricultural Gazette* to point out his priority, including some relevant passages from his 1831 book. In answer to this letter, Darwin himself acknowledged that Matthew had "anticipated by many years the explanation which I have offered of the origin of species, under the name of natural selection," and elsewhere he states, "Mr. Patrick Matthew . . . gives precisely the same view on the origin of species as that . . . propounded by Mr. Wallace and myself. . . . He clearly saw the full force of the principle of natural selection." Eventually, the somewhat curmudgeonly Matthew took to putting "The Discoverer of Natural Selection" on his calling cards.

Darwin communicated Matthew's claim to Wallace on May 18, 1860, writing, "Here is a curious thing, a Mr. Pat Matthew, a Scotchman, published in 1830 a work on Naval Timber and Arboriculture and in Appendix to this, he gives *most clearly* but very briefly in half-dozen paragraphs our view of natural selection. It is most complete case of anticipation" (emphasis Darwin's). Years later, after reading excerpts of Matthew's work in Samuel Butler's *Evolution, Old and New*, Wallace observed, "To my mind your quotations from Mr. Patrick Matthew are the most remarkable things in your whole book, because he appears to have anticipated the main ideas in *The Origin of Species*."

Since it is clear that both Darwin and Wallace admitted that Matthew was first off the mark, why is Matthew's contribution generally ignored in modern discussions of the origin of this crucial concept? Surely, the early thought processes of Matthew are just as relevant as those of Darwin or Wallace to the unraveling of the mystery of natural selection. This seems especially true when we consider Ernst Mayr's remarks that the concept of natural selection was

> so strange to Darwin's contemporaries when proposed in the *Origin of Species* in 1859 that only a handful adopted it. It took nearly three generations until it became universally accepted even among biologists. Among nonbiologists the idea is still rather unpopular, and even those who pay lip service to it often reveal by their comments that they do not fully understand the working of natural selection. It is only when one is aware of the complete unorthodoxy of this idea that one can appreciate Darwin's revolutionary intellectual achievement. And this poses a powerful riddle: How could Darwin have arrived at an idea which not only was totally at variance with the thinking of his own time but which was so complex that even now . . . it is widely misunderstood in spite of our vastly greater understanding of the processes of variation and inheritance?

It is even more puzzling to consider that Matthew, who had never traveled to exotic locales to collect specimens, as had Darwin and Wallace, and who was not part of a wide intellectual circle of correspondents, came to the same idea even earlier (in 1831), at a time when evolutionary concepts were even less popular or widespread. Mayr's characterization of the idea of natural selection as complex and totally at variance with contemporary thinking in 1859 only makes more paradoxical Matthew's statement that "to me the conception of this law of Nature came intuitively as a self-evident fact, almost without an effort of concentrated thought. . . . It was by a general glance at the scheme of Nature that I estimated this select production of species as an a priori recognizable fact—an axiom, requiring only to be pointed out to be admitted by unprejudiced minds of sufficient grasp."

How is it that this important piece of the evolutionary puzzle was discovered and yet largely ignored at the time? It may be argued that it was published in a place too obscure to be noticed by the scientific community. Mayr wrote elsewhere:

The person who has the soundest claim for priority in establishing a theory of evolution by natural selection is Patrick Matthew (1790–1874). . . . His views on evolution and natural selection were published in a number of notes in an appendix to his work *On Naval Timber and Arboriculture* (1831). These notes have virtually no relation to the subject matter of the book, and it is therefore not surprising that neither Darwin nor any other biologist had ever encountered them until Matthew brought forth his claims in an article in 1860 in the *Gardeners' Chronicle*.

Historian William Dempster has pointed out, however, that the subject of Matthew's book was the creation of improved stocks of trees for England's shipbuilders through artificial selection (a subject close to Darwin's heart). The contents of its appendix were reviewed in a popular journal of the time, by one of England's best-known botanists, and attracted enough notoriety for its unorthodox ideas to be banned from the local library. But perhaps it was an idea so far ahead of its time that it could not be connected to generally accepted knowledge and thus was consigned to the dustbin of premature and unappreciated ideas.

Evolutionary philosopher Daniel Dennett has criticized Matthew for relying on a simple deductive argument in formulating his law of natural selection. Yet the beauty of the idea, as with any powerful law in the natural sciences, is that it can be expressed in a few simple words or numbers. It was certainly this aspect of the idea that caused Thomas Henry Huxley to famously observe, "How stupid of me not to have thought of it."

But some have skirted the issue of Wallace's independent discovery by claiming that the most important contribution of Darwin was his profuse documentation of evolution and selection, published in *Origin of Species*. So we may well ask: Was the great achievement the idea or its elaboration and support? For me, the proposition of natural selection, clearly stated in a few paragraphs by Matthew, is more like a physical law. Once pointed out, the premises, concepts, deductions, and conclusions are straightforward.

Although Darwin admitted Matthew's priority, he downplayed the arboriculturist's contribution. He referenced Matthew's discovery in later editions of *Origin of Species*, but he added, "Unfortunately the view was given by Mr. Matthew, very briefly in scattered passages in an appendix to a work on a different subject, so that it remained unnoticed until Mr. Matthew himself drew attention to it in the *Gardeners' Chronicle* on April 7th, 1860."

The major criticisms that scientists have leveled against Matthew's ideas, however, revolve around his incorporation of the idea of catastrophic geologic change in his description of evolution by natural selection. Anthropologist Loren Eiseley concludes that Matthew's hypothesis remained obscure precisely because it was based on this "mistaken" cataclysmic view of the geologic record. Matthew cast his hypothesis in the mold of the then-current theories of Georges Cuvier and his followers, which, as we have seen, postulated a series of violent revolutions and successive inundations of Earth, during which organisms that existed during the previous epoch were destroyed. Paleontologist Martin Rudwick summarizes the status of geological studies at the time Matthew wrote his appendix:

By about 1830 . . . the spectacular success of some three or four decades of research on fossils had transformed Cuvier's early demonstration of a single recent organic revolution into a palaeontological synthesis of very wide scope and explanatory power. The geological time-scale was firmly established as almost unimaginably lengthy by the standards of human history, yet documented by an immensely thick succession of slowly deposited strata. The successive formations of strata . . . were clearly characterized by distinctive assemblages of fossil species, which enabled them to be identified and correlated over very wide areas. This correlation proved that in its broader outlines the history of life had been the same in all parts of the world.

Cuvier's recent "revolution" had turned out to be only the last of many similar events punctuating the history of life, and having apparently sudden and drastic effects not only on the terrestrial faunas that he had reconstructed, but also on the more abundant marine faunas and on plant life as well. These revolutions were evidently natural events that were somehow built into the physical constitution of the globe, and in character they appeared to be sudden transitory tidal waves that swept over at least the low-lying areas of the continents. Their physical cause remained uncertain. . . . In any case, as drastic episodes of environmental change they were adequate to explain the mass extinction of faunas and floras of well-adapted species.

Matthew was well aware of the fossil record of extinctions and favored a naturalistic explanation for episodic mass extinctions:

A particular conformity, each after its own kind, when in a state of nature, termed species, no doubt exists to a considerable degree. . . . Geologists discover a like particular conformity—fossil species—through the deep deposition of each great epoch, but they also discover an almost complete difference to exist between the species or stamp of life of one epoch from that of every other. We are therefore led to admit, either of a repeated miraculous creation; or of a power of change, under a change of circumstances, to belong to living organized matter. . . . The derangements and changes in organized existence, induced by a change of circumstance . . . affording us proof of the plastic quality of superior life, and the likelihood that circumstances have been very different in the different epochs, though steady in each, tend strongly to heighten the probability of the latter theory.

And for Matthew, catastrophic mass extinctions were critical to the process of evolution:

The destructive liquid currents, before which the hardest mountains have been swept and comminuted into gravel, sand, and mud, which intervened between and divided these epochs, probably extending over the whole surface of the globe, and destroying nearly all living things, must have reduced existence so much, that an unoccupied field would be formed for new diverging ramifications of life. . . . [T]hese remnants, in the course of time, moulding and accommodating their being anew to the change of circumstances, and to every possible means of subsistence, and the millions of ages of regularity which appear to have followed between the epochs, probably after this accommodation was completed, affording fossil deposit of regular specific character.

Thus Matthew could explain both extinctions and the subsequent radiations of species in his catastrophic framework. Stephen Jay Gould at Harvard recognized that, by contrast, mass extinctions were "a crucial issue that caused Darwin no end of trouble," from which he "could extract himself only by claiming that mass extinctions were artifacts of an imperfect rock record."

Unfortunately, Matthew published his ideas, tied as they were to a catastrophist interpretation of the geologic record, just as Lyell brought out his monumental *Principles of Geology*, espousing the gradualistic and non-catastrophic view of Earth history. Lyell triumphantly proclaimed, "All theories are rejected which involve the assumption of sudden and violent catastrophes and revolutions of the whole earth, and its inhabitants." So convincing were Lyell's arguments for his gradual picture of geologic change that they soon supplanted Cuvier's catastrophism with an orderly and non-revolutionary model of the geologic record that, as we have seen, colors our geologic interpretations to the present day.

But the geological tides have turned, and research over the past 35 years, beginning with the Alvarez hypothesis in 1980, has shown that mass extinctions are a very real phenomenon in the history of life, and catastrophic causes, both terrestrial and extraterrestrial, are prime suspects in several cases. If mass extinctions are a critical factor in evolution, then Darwin's dismissal of these events as artifacts of the geologic record is a significant oversight. Changes in our interpretation of the geologic record and of the mode and tempo of evolution demand a reconsideration of Matthew's contribution. His ideas may be a more accurate representation of evolutionary processes than those unveiled by Darwin and Wallace some 27 years later.

Today, the geologic record is seen as composed of long intervals of relative stasis of fauna and flora, terminated by extinction events apparently caused by severe disruptions of the physical environment. Habitat fragmentation is widely assumed to be essential for speciation, and the presence of refugia during cataclysmic extinction events would provide for extensive fragmentation of terrestrial and marine habitats. Thus the idea that most evolutionary change was accomplished very gradually, through competition between organisms and through organisms becoming better adapted to relatively stable environments, is being supplanted by a recognition that major morphological and ecological changes tend to occur episodically and rapidly.

When the origins of the theory of evolution by natural selection are discussed, Darwin is always given the dominant role, with Wallace mostly an afterthought. However, Matthew deserves to be acknowledged for his earlier formulation of natural selection and, more specifically, for his much more accurate assessment of the catastrophic nature of the geologic record and the origin of species. Gould, for one, concluded that, despite the clear priority of Matthew's discovery of natural selection, he deserves little credit because he missed the larger significance of his own brainchild. I beg to differ. In a prescient foreshadowing of Darwin's famous closing statement in *Origin of Species* ("From so simple a beginning endless forms most beautiful and most wonderful have been, and are being, evolved"), Matthew reveals a clear grasp of the aesthetics and power of this view of life:

> Does organized existence, and perhaps all material existence, consist of one . . . principle of life, capable of gradual circumstance-suited modifications and aggregations, without bound . . . ? There is more beauty and unity of design in this continual balancing of life to circumstance, and greater conformity to those dispositions of nature which are manifest to us, than in total destruction and new creation.

7

• • • •

Impacts and Extinctions

DO THEY MATCH UP?

An immediate question that arose from the Alvarez et al. discovery
was whether similar events were associated with one or more of
the other mass extinctions.

CHARLES ORTH, "GEOCHEMISTRY OF THE BIO-EVENT HORIZONS"

Doubters and gradualists remain, but the results are clear. The sudden end-Cretaceous extinctions were caused by a cosmic impact. The impact hypothesis has met every reasonable test, and it is clear from location, geochemistry, and radiometric dating that the Chicxulub impact is the culprit. The discovery of the end-Cretaceous impact layer prompted a search for impact signatures at other geologic boundaries marked by significant extinction events. Scientists worldwide carried several diagnostic strategies with them. In addition to noting elevated concentrations of iridium and related trace elements in cosmic proportions, their field and laboratory work sought other impact materials: shocked minerals (including shocked quartz, feldspar, and zircon); high-density phases of quartz (coesite and stishovite); and impact glass (tektites and microtektites). Microspherules of various mineral compositions, showing crystallization at high temperatures and including nickel-rich crystals of the mineral spinel, are also diagnostic of impact. Furthermore, they considered coarse near-source deposits, mostly mass-flow deposits of various kinds, plus related tsunami-induced sediments from large oceanic impacts.

Biological and geochemical characteristics of the Cretaceous/Paleogene boundary that might exist at other mass-extinction boundaries include (1) a globally synchronous or near-synchronous mass-mortality level, marked by a shift in carbon isotope ratios in marine sediments, indicating a catastrophic loss of biomass and a drop in ocean productivity (the so-called Strangelove ocean); (2) a proliferation of disaster species and opportunistic organisms, followed by a recovery and radiation of surviving species; (3) a brief cooling; (4) a marked shift in oxygen isotope ratios, suggesting a longer-term global warming; and (5) a sharp reduction in marine calcium carbonate produced by shelled organisms (see figure 5.5).

Many geologists seem to think that a concerted effort has been made to search carefully for evidence of impacts at other geologic extinction boundaries and that the results of these searches have been entirely negative. Nothing could be further from the truth. Searching for impact evidence in the geologic record has been sporadic and is not easy. Impact layers are typically only a few millimeters to centimeters thick and occur relatively rarely in the record—finding them is equivalent to finding a needle in a very large haystack (figure 7.1).

FIGURE 7.1 Impact layers are like needles in a haystack. The Vispri Quarry near Gubbio, Italy, represents more than 40 million years of continuous deposition in a deep marine environment during the Cretaceous and Paleogene periods. Sedimentary layers are tilted down to the left. The truck on the right shows the scale.

At the time of the discovery of the Chicxulub crater, in 1991, only a few, relatively recent stratigraphic levels had yielded evidence of impact. For example, several well-known tektite/microtektite layers date from the past 2 to 3 million years: the Australasian tektites/microtektites, largely in the eastern Indian Ocean and surrounding land areas (780,000 years old), possibly from a source crater at Tonlé Sap Lake in Cambodia; the Ivory Coast tektites/microtektites in West Africa and the adjacent Atlantic Ocean (1.1 million years old), from the Bosumtwi crater; and melted debris left on the ocean floor by the Eltanin event (2.3 million years ago), a small impact in the southern Pacific Ocean, though the expected small crater on the seafloor has not yet been found.

Older examples are the moldavite tektites in the Czech Republic (15 million years old), from the Ries crater in Germany, and the North American tektites/microtektites in the southeastern United States and adjacent ocean (36 million years old), now known to be products of the Chesapeake Bay impact structure. In 2003, I traveled with noted geochemist Christian Koeberl of the University of Vienna to the Western Desert of Egypt to collect Libyan desert glass (29 million years old), which is the result of an undiscovered impact crater. We camped out for a week in the trackless desert and collected numerous samples of the impact-melted glass for analysis. Back in the lab, Koeberl discovered the signature of extraterrestrial matter in dark bands within the yellowish glass.

Many of the searches for impact fallout deposits may not have sampled closely enough to detect very thin impact layers. First, one must find the layer representing the exact timing of the mass extinction, which is often difficult due to incomplete sampling and to geologic processes, such as sediment reworking and burrowing organisms, that tend to smear out the record of extinctions over time. Even if your search for an impact layer gives negative results, you can't be sure that such a thin layer wasn't present. You may easily have missed it in your sampling protocol. Sampling should be continuous, wherever possible, but this is very time consuming and requires a lot of lab work. Sampling every 10 or 20 centimeters (4 or 8 inches), as is commonly done, may not be sufficient to detect thin distal impact layers.

Furthermore, iridium and shocked quartz in the sediments occur in amounts of parts per billion (and sometimes less), and parts per billion are difficult to detect. One must take very large samples if one is to find the very rare shocked grains and microtektites, the abundance of which may be a few grains per cubic centimeter of sample. But the samples must also be close together so that you don't miss the thin impact layer—something that is difficult to do in practice. A cursory look at the stratigraphic record usually will not reveal the impact evidence.

The Cretaceous/Paleogene boundary has been used as the standard of reference in the search for impact-related iridium anomalies. As we have seen, this iridium anomaly, which is in the range of several parts per billion (tens to hundreds of times the normal background iridium level in sediments), is indeed globally well documented at that boundary. But Frank Kyte of UCLA and others have argued that the high iridium concentration at the Cretaceous/Paleogene boundary should perhaps not be considered typical of stratigraphic iridium anomalies from impacts. For example, the unusual target rock (a shallow-water carbonate and evaporite platform) created an expanding volatile cloud that probably affected the composition and distribution of ejected material. Some geologic boundaries show only a few hundred parts per trillion of iridium, including horizons known to be associated with large impact-cratering events like the Late Eocene Popigai crater and its impact debris layer, so this amount might be enough to suggest impact. If impact is suspected, then study of other trace elements and isotopes should be used to test for a meteoritic source, and shocked quartz and microtektites should be searched for. Also, remember that comets may be mostly ice and therefore might leave only small iridium anomalies.

Two additional factors, however, make it especially problematic to accurately estimate impact crater size from stratigraphic iridium anomalies: there is a great range of iridium content in different types of extraterrestrial bodies, and large impacts tend to lose ejected material through atmospheric blowoff. First, iridium concentrations in known meteorites vary, from 30 parts per trillion in rocky achondrite and eucrite meteorites to 100 parts per million in some iron meteorites. Walter Alvarez and his team originally argued for a primitive undifferentiated carbonaceous chondrite composition for the end-Cretaceous impactor, but others have suggested that element-abundance patterns at the boundary could be used to argue for various impactors, ranging from metal sulfide objects to comets. More recently, however, chromium isotope analysis of the Cretaceous/Paleogene boundary clay supports a carbonaceous chondrite type of impactor. Some impact specialists believe that most large craters are produced by comets, and the non-ice portions of comets are very likely to have a primitive carbonaceous chondrite composition.

Second, planetary scientists Ann Vickery and Jay Melosh at Purdue calculated iridium loss through atmospheric blowoff. Their atmospheric model predicts that, among impacts producing the largest craters, blowoff of the atmosphere could cause significant iridium loss from Earth. Impacts of large, ice-rich, fast-moving long-period comets, capable of causing severe environmental disturbances, might therefore produce only relatively weak iridium spikes in the geologic record. This may explain abrupt mass-extinction events that seem to lack distinct iridium anomalies.

Using the information gained from study of the Cretaceous/Paleogene boundary, we can look for evidence of impact in other parts of the geologic record, but there are several problems. First, where exactly should we look? Fortunately, in the case of the end-Cretaceous event, the boundary at Gubbio and elsewhere is well marked by the extinction of most of the calcareous plankton, at the thin layer of clay. The boundary layer was thus readily found, even though the layer looks pretty much the same as many other clay bands in the Gubbio outcrop. It was discovered because of its special relationship to the well-marked severe mass-extinction event. The correlation with impact at other extinction boundaries, where sediment mixing has taken place or where fossils are scarce, might not be so evident.

As of this writing, potential impact evidence, including elevated iridium, microtektites, microspherules, or shocked minerals, has been reported at or near at least six extinction events (see table 5.1). We have explored the end-Cretaceous event in detail. We'll take a close look at each of the others in turn. But remember, these occurrences are often ignored or explained away as a result of noncatastrophic processes. Some geologists have claimed repeatedly that the only bona fide incidence of impact debris at a geologic boundary is the Cretaceous/Paleogene event. They seem reluctant to go beyond the one-off impact event at the end of the Cretaceous, even when there is evidence of impact. I think this is another example of the effects of Charles Lyell's laws on modern geology.

The six key events are

1. The end-Cretaceous (66 million years ago), with iridium, shocked quartz, and microtektites/microspherules
2. The Late Eocene (36 million years ago), with iridium, shocked quartz, microtektites, and microspherules
3. The Jurassic/Cretaceous boundary (145 million years ago), with iridium, shocked quartz, and microspherules
4. The Bajocian/Bathonian boundary (168 million years ago), with elevated iridium and microspherules
5. The Middle Norian (215 million years ago), with iridium, shocked quartz, and microspherules
6. The Late Devonian (372 million years ago), with some iridium and multiple microtektite layers

What's more, each of these six events is associated with large impact craters (figure 7.2).

The Late Eocene

The Alvarez team was the first to search for iridium anomalies at other boundaries. A relatively small extinction event (about 30 percent species extinction, according to David Raup and Jack Sepkoski) occurred during the Late Eocene period, about 36 million years ago. Other researchers had already discovered one or more layers of microtektites near that geologic boundary, and some suggested that comet impacts might be responsible for the geologic discontinuity. In 1982, the Berkeley group obtained samples from an Ocean Drilling Program drill core, while Ramachandran Ganapathy of the J. T. Baker Chemical Company in New Jersey independently examined samples from an ocean core collected by the Lamont-Doherty Earth Observatory.

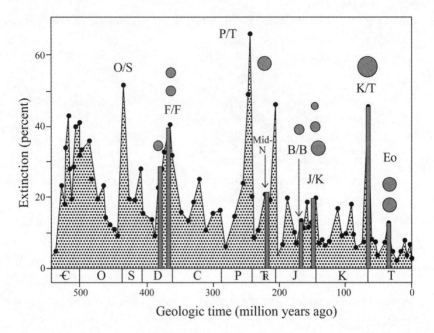

FIGURE 7.2 Extinction events showing stratigraphic evidence of impact and correlation with large-impact craters. Circles are dated large-impact craters; columns are geologic boundaries with evidence of impacts.

The Berkeley scientists and Ganapathy published their results side by side in *Science*. Nuclear activation analyses near one of the microtektite-rich layers showed anomalous concentrations of iridium—4 parts per billion in Ganapathy's sample but only 300 parts per trillion in the Alvarez group's sample, because they missed the peak value due to a gap in their core. This raises the question of how many other relatively small iridium anomalies are "small" only because the peak values were missed in sampling. The iridium anomaly in Ganapathy's core is about 30 centimeters (12 inches) below a microtektite layer, so these apparently represent two separate impact events.

The standard geologic section for the Late Eocene is the Massignano outcrop in Italy, near the city of Ancona on the Adriatic Sea, where the boundary has been precisely dated by radiometric and rock-magnetic methods (figure 7.3). At Massignano, three iridium peaks dating to the Late Eocene can be found, but in amounts less than one-tenth of the Cretaceous/Paleogene anomaly (though perhaps only the shoulders of the anomalies were sampled). In 1996, however, geology student Aaron Clymer of Carleton College and co-workers reported the discovery of shocked quartz at the lowermost iridium peak, with a frequency of only one to two grains per cubic centimeter of bulk rock (figure 7.4). The discovery of another signature of impact came in 1998. Olivier Pierraud of the Centre des Faibles Radioactivities in France, Allesandro Montanari of the Geological Observatory of Coldigioco in Italy, and co-workers found nickel-rich crystals of the mineral spinel embedded in flattened microspherules at the same level in the rocks. The microspherules found at Massignano have been radiometrically dated (using nearby volcanic ash layers) at about 36 million years ago and thus correlate quite well with the age of the 100-kilometer-diameter (62-mile) Popigai impact crater in Siberia.

Ken Farley of Caltech (an expert on detecting extraterrestrial signals in ancient rocks) provided another feature related to the impacts. Farley found an increase in extraterrestrial helium-3 in an Upper Eocene ocean core from the North Pacific. He interpreted this increase, which occurred over the span of a few million years, as the result of an influx of interplanetary dust

FIGURE 7.3 Gene Shoemaker at the section of Late Eocene rocks in Massignano, Italy, where iridium, shocked quartz, and microspherules were found.

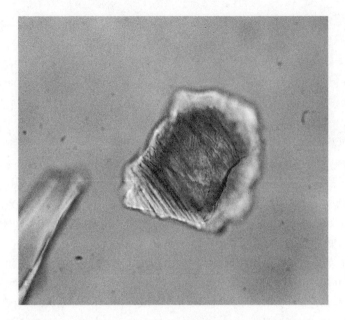

FIGURE 7.4 A shocked quartz grain from the impact layer at Massignano, showing multiple directions of planar deformation features. The grain is about 100 microns (0.0039 inch) in diameter.

particles, possibly shed from increased numbers of comets (a comet shower) or asteroids passing through the inner solar system. In the Massignano outcrop, Farley, working with Eugene Shoemaker and Montanari, found that the interval of enhanced helium-3 covered a span of about 2.2 million years, and it includes the three iridium anomalies and the impact debris layer. The uppermost iridium spike coincides with a helium-3 peak about six times higher than background levels, making it an interesting candidate for further analyses of impact signatures. I have looked for shocked quartz at this level with no success thus far, but my sampling may not have been dense enough, or the samples large enough, to find the tiny grains.

Biotic changes seem to have begun in the Middle Eocene and, according to Sepkoski's data, marine species apparently suffered an extinction ranging from the Late Eocene through the Early Oligocene epochs. Six species of planktonic radiolarians went extinct at that time. According to Gerta Keller at Princeton, the microtektite event (or events) seems to be associated with 1 to 2 million years of stepwise events of faunal turnover in calcareous plankton, with a total extinction involving about 60 percent of the plankton population. Rodolfo Coccioni, a micropaleontologist at the University of Urbino, and his colleagues found no abrupt extinctions of plankton at the Popigai impact layer at Massignano, but significant changes in calcareous plankton occurred about 60,000 years after the impact layer, where one might expect microtektites from the Chesapeake Bay crater. But Simonetta Monechi of the University of Florence, an expert on tiny calcareous ocean plankton, found a marked change in the fossil assemblage across the Popigai impact event layer.

The Jurassic/Cretaceous Boundary

According to Sepkoski's data compilation, the Jurassic/Cretaceous transition at about 145 million years ago was marked by the disappearance of about 40 percent of marine species. Anthony Hallam, a noted geologist at the University of Birmingham, has emphasized that mollusks show a drastic reduction during Late Jurassic times, although the mollusk extinction seems to have been concentrated regionally in western Europe and Russia. Henning Dypvik of the University of Oslo traced a layer of ejected material from the 40-kilometer-diameter (25-mile) Mjølnir impact structure (dated at about 143 million years ago) in the Barents Sea to Upper Jurassic/ Lower Cretaceous rocks in a series of drill holes off northern Norway, more than 800 kilometers (500 miles) from the impact. The layer contains shocked quartz grains and an iridium anomaly of up to about 1 part per billion, with element ratios similar to that of meteorites. It is placed very close to the regional Jurassic/Cretaceous boundary in the Northern Hemisphere (recently dated at around 143 million years ago).

This agrees with a large iridium anomaly (up to 7 parts per billion) in a Jurassic/Cretaceous sequence in northern Siberia, reported by geologist Andre Zhakarov, an expert on Jurassic/Cretaceous rocks, and his Russian colleagues. The iridium anomaly, with element ratios similar to that of primitive chondritic meteorites, and abundant chemically altered spherules made of the mineral pyrite (iron sulfide) occur in a thin limestone layer at the same stratigraphic level as the layer of ejected material near the

Mjølnir impact, some 2,300 kilometers (1,400 miles) distant. Thus although the worldwide formal correlation of the Jurassic/Cretaceous boundary remains unsettled, there is now evidence that the transition between those two periods, in the Northern Hemisphere, was associated with the widespread ejecta layer from a 40-kilometer-diameter (25-mile) complex impact structure.

In England, the correlative horizon lies within the nonmarine Purbeck Beds of the uppermost Jurassic strata, at an exotic, oyster-rich marine intercalation called the Cinder Bed. Scientists also have described a coarse deposit from a similar stratigraphic level in France, possibly the result of a tsunami. These European sites have been examined for iridium, with negative results thus far. The lack of an international definition of the Jurassic/Cretaceous boundary stems in part from the difficulty of correlating northern faunas with those of the equatorial Tethyan region. In the Tethyan realm, the Jurassic/Cretaceous boundary has been set at a slightly older level of faunal change, at about 145 million years ago.

At the beautifully exposed geologic section that includes the Jurassic/Cretaceous boundary in the Bosso Gorge in central Italy (figure 7.5), Koeberl and his students at the University of Vienna found several iridium spikes of between 100 and 200 parts per trillion. These spikes are below the parts per billion cutoff used by most geologists, based on the iridium spike at the Cretaceous/Paleogene boundary. But the end-Jurassic anomalies are well above background iridium concentrations. Furthermore, in the Bosso section, one of the layers enriched in iridium is unusual and appears to have been formed by a rapidly moving current originating outside the area (perhaps correlative with the deposits in northern Europe), possibly produced by the Mjølnir impact. This would be a good place to look for microtektites and shocked minerals.

In the Southern Hemisphere, the presence of two impacts—the well-dated Gosses Bluff crater in Australia (22 kilometers [14 miles] in diameter and about 143 million years old) and the large Morokweng structure in South Africa (at least 70 kilometers [43 miles] in diameter and about 145 million years old)—suggests that several impacts and faunal turnover events, over a period of several million years, may have been involved in the global Jurassic/Cretaceous transition. These multiple impacts could

FIGURE 7.5 The Jurassic/Cretaceous boundary at Bosso Gorge in Italy. The boundary is approximately at the hand of the kneeling figure of Sandro Montanari.

explain the regional nature of some of the Jurassic/Cretaceous extinctions and the problems in defining and correlating a formal global stratigraphic boundary.

The Bajocian/Bathonian Boundary

The 80-kilometer-diameter (50-mile) Puchezh-Katunki crater in Russia recently has been dated at about 167 million years ago. It is thus correlative with the latest age of the Bajocian/Bathonian boundary in the Middle Jurassic, at 168 million years ago. Raup and Sepkoski list this boundary as a doubtful extinction event at the family level but as a minor extinction peak in their genera-level data. Geochemist Robert Rocchia and his co-workers from the University of Paris found an iridium anomaly of 3.2 parts per billion, in a layer a few millimeters thick, in a Bajocian/Bathonian marine sequence in northern Italy. Microspherules were later discovered in the same layer. The iridium anomaly and the microspherules were interpreted by Rocchia and his colleagues to be the result of concentration of the element by oceanic processes, but an impact origin also is a strong possibility. Two other, smaller impact craters in Russia—Zapadnaya, about 166 million years old, and Obolon, about 169 million years old—may also have contributed to the end-Bajocian event.

The Middle Norian

The large, 100-kilometer-diameter (62-mile) Manicouagan impact structure in Quebec has a currently accepted radiometric age of about 214 million years, which correlates with the Middle Norian stage in the Late Triassic. At that time, a number of reptile genera and flora in the western United States died out, and a catastrophic collapse of the ocean ecosystem has been reported from Japan. Most significantly, an impact-related layer containing iridium and shocked quartz, dating to about 215 million years ago and interpreted as having come from the Manicouagan impact, occurs in Upper Triassic deposits at widely separated sites—in the British Isles and, on the other side of the world, in Japan.

The Late Devonian

The first iridium peak discovered to correspond to the Late Devonian mass extinction (the Frasnian/Famennian stage boundary), about 372 million years ago, was found in the Canning Basin in Australia, where it occurs within the fossilized remains of calcareous algae. The anomaly has been explained away as a potential biological concentration of the iridium and other trace elements. But similar occurrences of algae above and below the boundary contain no excess iridium. Why should iridium concentrations peak only near the extinction boundary? Perhaps the algae simply concentrated the element from iridium-enriched waters as a result of an impact. Such microbial limestones are found in connection with several mass extinctions and seem to represent the spread of opportunistic algal species in the absence of predators that disappeared in the mass extinctions. Microtektites and shocked quartz were not looked for at the iridium anomaly.

Study of Upper Devonian deposits (including the Frasnian/Famennian stage boundary) in China by Kun Wang and his colleagues at the University of Alberta reported somewhat elevated iridium concentrations of 230 parts per trillion. This still is an order of magnitude less than in the Cretaceous/Paleogene boundary clay, but it is almost 15 times the background concentration in the rocks above and below the boundary. Similarly, George McGhee Jr. at Rutgers and his colleagues searched in New York State and in Belgium for geochemical evidence of impact at the time of the Late Devonian mass extinction and found only relatively low levels of iridium (up to 120 parts per trillion), but they undertook no intensive searches for shocked quartz or microtektites.

Several small iridium peaks (around 100 parts per trillion over a background of only 20 to 30 parts per trillion) from the same Late Devonian interval have been reported at La Serre, in southern France, but microtektites were not searched for. At Xiangtian, Guangxi, in southern China, a Late Devonian carbon isotope shift correlates with an iridium anomaly of about 350 parts per trillion over a background of only around 30 parts per trillion but, again, no hunts for microtektites or shocked quartz were undertaken at the iridium anomaly.

Specific searches for microtektites have been more successful. Chinese scientists X. P. Ma and S. L. Bai from Beijing University found several microtektite layers near the extinction level of marine species in southern China (figure 7.6), and similar microtektites, from the same Upper Devonian extinction level, were reported in Belgium by Philippe Claeys and his colleagues at the Free University of Brussels (figure 7.7). The Claeys group found a small iridium anomaly exactly in the microtektite layer, with amounts above background levels but, again, considerably weaker than that found at the Cretaceous/Paleogene boundary.

Remember, however, that the Late Eocene iridium anomalies at Massignano were only a few hundred parts per trillion, and we know that they

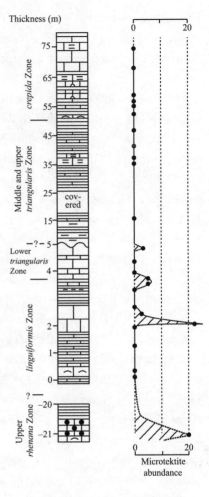

FIGURE 7.6 A section across the Frasnian/ Famennian (Late Devonian) boundary (close to the base of the lower triangularis conodont zone) in Xikuangshan, China. Note the three or four layers with microtektites. Numbers are microtektites per hundred grams of sediment.

were associated with microtektites, microspherules, shocked quartz, and two large impacts. Furthermore, the sampling protocol of Claeys's group in the Frasnian/Famennian boundary zone was only one sample every 5 centimeters (2 inches), which means that they may have missed a thin peak iridium layer. Another possibility is that the weak iridium peaks are the result of the impact of ice-rich comets containing relatively small amounts of iridium.

It seems clear that one or more microtektite horizons occur in the Late Devonian, but they are seemingly accompanied by only small iridium

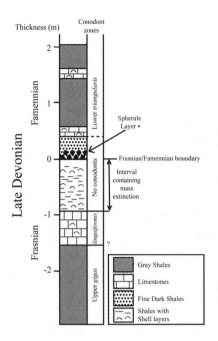

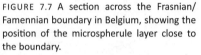

FIGURE 7.7 A section across the Frasnian/Famennian boundary in Belgium, showing the position of the microspherule layer close to the boundary.

anomalies—just what one would expect from the impacts in a comet shower. Kun Wang and Helmut Geldsetzer of the Geological Survey of Canada reported a 20-centimeter-thick (8-inch) "storm" bed in Alberta, Canada, exactly at the Frasnian/Famennian boundary, which could be an impact-related tsunami deposit.

Then there is direct evidence of impacts. Four reasonably well-dated impact craters (Siljan, Woodleigh, Ilyinets, and Kaluga) have radiometric ages that overlap the estimated age of the Frasnian/Famennian extinction boundary, which was recently revised to about 372 million years ago. The relatively large, 55-kilometer-diameter (34-mile) Siljan impact structure in Sweden has been dated at about 377 million years ago. Twenty-two-kilometer-wide (14-mile) Lake Taihu in southeastern China, near Shanghai, also may be the remains of a larger Late Devonian impact crater (figure 7.8). A somewhat earlier impact event (about 382 million years ago) produced the impressive Alamo Breccia deposits in southwestern Nevada, where shocked quartz and microspherules have been discovered. It seems that the rate of impacts in the Late Devonian may have

FIGURE 7.8 The 22-kilometer-diameter (14-mile) Lake Taihu impact structure in China, which may be of Late Devonian age.

been unusually high for millions of years (see figure 4.1). An impactor shower is a possibility, perhaps a rain of comets or an asteroid surge that led to increased rates of extinction throughout the 30 million years of the Late Devonian.

8

• • • •

The Great Dying

THE END-PERMIAN EXTINCTIONS

The new world that began with the recovery from the end-Permian
extinction was utterly different from the world of the Paleozoic.

DOUGLAS ERWIN, *EXTINCTION*

The mass-extinction event at the end of the Permian period, 252 million
years ago, is the most severe mass extinction in the geologic record, and
evidence of impact would give much credence to the linkage between
impacts and mass extinctions. The event is marked by the disappearance
of up to 96 percent of marine species, the extinction of more than 90
percent of reptile genera, a severe loss of insects, and a major floral extinction, including the abrupt obliteration of the widespread *Glossopteris* flora
on the southern continents. The extinction was followed by the explosion
of opportunistic species, such as marine acritarchs (small noncalcareous
plankton) in the ocean and fungi, ferns, and primitive club mosses on
land. This suggests a sudden global catastrophe in the sea and on land, followed by a "disaster ecology" of unusual species. Planktonic and nektonic
groups and groups with planktonic growth stages show a sudden crash,
and the deposition of calcium carbonate from shelly fossils decreases in
many sections, replaced by opportunistic cyanobacterial encrustations and
inorganic precipitation of carbonates directly from seawater (figure 8.1).

Significant shifts in the isotopes of carbon, oxygen, and sulfur also mark
the end-Permian event. The extinction seems to be linked to a time of
extremely high temperatures; oxygen isotope studies of early Triassic deposits have been interpreted as indicating ocean surface temperatures of up
to an extreme of 38°C (100°F). It has even been suggested that tropical
areas were so hot as to be unsuitable for many kinds of life.

FIGURE 8.1 The Permian/Triassic boundary at Seres in the Italian Alps. The mass-extinction level is the thin dark layer in the limestone just above the head of the seated figure (Paul Wignall). This dark layer may indicate a lack of dissolved oxygen in the upper layers of the ocean at that time.

Is there any impact evidence at the boundary? Weak iridium peaks in the hundreds of parts per trillion range were reported from boundary sections in Austria, China, India, and Italy. In the Austrian Alps, a double peak in iridium correlates with major carbon isotope shifts, but the iridium anomalies show geochemical evidence of an earthbound source, and a volcanic origin has been suggested (figures 8.2 and 8.3). At least one impact is known to have occurred around the time of the Permian/Triassic boundary: the 40-kilometer-wide (25-mile) Araguainha impact structure in Brazil has a radiometric age of about 255 million years ago. I studied this remarkable impact structure on the ground in Brazil in the summer of 2000. It is the largest recognized impact structure in South America, but most impact specialists feel that it is too small to have caused the end-Permian global destruction.

In the same year, I also collected samples from Japanese outcrops at the Permian/Triassic boundary and worked with geochemist Luann Becker from the University of California, Santa Barbara, and her colleagues to analyze samples of the boundary from China and Japan. In 2001, Becker and

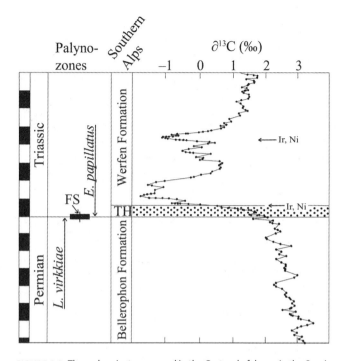

FIGURE 8.2 The carbon isotope record in the Gartnerkofel core in the Carnic Alps in Austria. Note the two iridium/nickel anomalies at the two major decreases in the carbon isotope ratios, and the occurrence of the fungal spike (FS) in the pollen/spore record, which is projected from other localities. White and dark bands on the left are about 100,000 years long. The Tesoro horizon (TH) is an unusual sedimentary unit that occurs close to the Permian/Triassic boundary in many localities in the Alps.

her co-workers reported a peak of the substance fullerene at the Permian/Triassic boundary. These fullerenes are carbon-60 molecules, with the 60 carbon atoms arranged in a "soccer ball" structure reminiscent of the geodesic domes built by the inventor Buckminster Fuller, hence the name fullerene. Becker and coauthor Robert Poreda from the University of Rochester claimed that they had detected rare gases, including helium-3, in meteoritic proportions, trapped within the cages of the fullerene molecules. This would suggest that the fullerenes had their source in an asteroid or a comet impact.

Unfortunately, no one has been able to repeat the analysis. Ken Farley of Caltech looked for extraterrestrial helium-3 in samples from the boundary

FIGURE 8.3 The author at the Permian/Triassic boundary at the Gartnerkofel in the Austrian Carnic Alps. I am sitting on the Permian/Triassic
boundary, with Permian rocks below me and Triassic rocks above.

in China and couldn't find any. Becker claims that Farley missed the
impact layer. She analyzed a dark, pyrite-rich (iron sulfide) layer, whereas
Farley may have looked at the overlying volcanic ash layer, which would
not have any fullerenes or helium-3. So the discovery of impact-related
fullerenes at the Permian/Triassic boundary is still not confirmed. Becker
and her co-workers also brought attention to a possible end-Permian impact
crater, the Bedout structure, off the western coast of Australia. But the

evidence cited by Becker's group in identifying this structure as related to an impact, and its Permian/Triassic age, has not held up to scrutiny; no unambiguous shocked material was found in drill holes into the structure, and the breccia deposits look volcanic in origin.

The world's standard section for the Permian/Triassic boundary is in an outcropping of rock near the small village of Meishan in southeastern China (figure 8.4). At the site, the Chinese have developed an elaborate science-information center, celebrating the presence of the global type section for the boundary. The outcrops occur in an abandoned limestone quarry, and the rocks are a mixture of marine limestones and mudstones. The Permian/Triassic boundary has been placed within one of the limestone beds (bed 27), based on the first appearance of the earliest typical Triassic fossils. The mass-extinction level, however, is a little below this (at the base of bed 25), at a layer rich in pyrite.

A similar dark, pyrite-rich layer is present in a number of Permian/Triassic sections around the world. It may represent a time of sulfide deposition during a period of low ocean oxygenation, even in shallow ocean

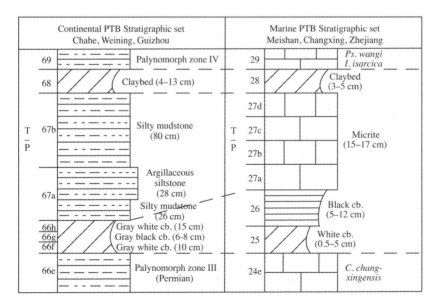

FIGURE 8.4 Two Permian/Triassic sections in China—the terrestrial section at Chahe (*left*) and the marine section at Meishan (*right*)—each showing two well-dated volcanic ash beds (*diagonal lines*) widely distributed in southern China.

waters. This is supported by the presence in end-Permian sediments of organic chemicals derived only from photosynthetic green sulfur bacteria. These bacteria would have prospered in an ocean with anoxic surface waters. Later, we will come back to the potential correlation between extinctions and anoxic periods in the ocean.

The section at Meishan is also important for determining Late Permian and Early Triassic timescales, because it has been found to contain several beds of altered volcanic ash that span the boundary (beds 25 and 28; see figure 8.4) and are widespread in both marine and nonmarine sequences in southern China. The presence of these ash layers across more than 1 million square kilometers (386,000 square miles) in southern China indicates several explosive super-eruptions during the time of the extinction. It may be that these eruptions contributed to environmental changes at that time. At Meishan, these layers have been repeatedly dated radiometrically, with growing accuracy, giving an age of just over 252 million years for the extinction level. One way to establish the rate of the extinctions is to utilize these dated ash beds to calculate sediment accumulation rates at the time of the mass die-off.

According to Kunio Kaiho, a geochemist at Tohoku University in Japan, the extinction at Meishan occurs abruptly within a bed that is only 1.2 centimeters (about 0.5 inch) thick, essentially a single bedding plane near the top of bed 24 (see figure 8.4). If we use the accumulation rate determined for the Meishan sediments (a very low rate of about 0.4 centimeter [0.16 inch] per thousand years), then the extinction layer would represent only about 3,000 years. Better resolution of the Permian/Triassic boundary interval can be found in deposits at Gartnerkofel in the Carnic Alps of Austria (see figure 8.3). Unfortunately, no datable ash layers have been found near the boundary in that section, but estimates of deposition rate, based on the presence of known astronomical cycles of Earth's orbit, which are recorded as changes in sedimentation in the rocks, put the entire transition, from the extinction event to the first appearance of typical Triassic fossils, all within less than 8,000 years. Surely, these results demand some cataclysm.

In many Permian/Triassic boundary sections, the end-Permian extinctions can be correlated with a pronounced anomaly in carbon isotopes (see figure 8.2). Ken Caldeira and I attributed this anomaly to a shutdown in ocean productivity, which led to less partitioning of carbon between shallow and deep waters and hence an increase in light carbon-12 in surface waters—the so-called Strangelove ocean effect, which I discussed in relation to the Cretaceous/Paleogene boundary. Another possibility is the addition of light carbon to the ocean–atmosphere system by increased volcanism, soil erosion, or a large-scale release of methane, normally stored in continental shelf sediments. Deposits produced by calcareous algae and nonbiogenic calcium carbonate deposited directly from warm ocean waters are also characteristic of the Permian/Triassic boundary. These may have been the result of increased dissolved calcium carbonate in the waters, as very little carbonate was being deposited by the decimated marine organisms. As the oceans became more saturated in calcium carbonate, nonbiological deposition would come to dominate.

Scientists also use another method of locating the Permian/Triassic boundary, taking advantage of the "fungal event" or "fungal spike" that appears in many Permian/Triassic boundary sections, both marine and nonmarine, around the world (see figure 8.2). (A fungal spike also occurs at the Cretaceous/Paleogene boundary.) This fungal event, believed to represent a global event time line, follows the destruction of Late Permian forests worldwide and is most likely the result of proliferation of fungi on the rotting vegetation produced by forest die-out, with fungal material and woody debris accounting for more than 90 percent of the pollen and spores in the aftermath of the forest plant extinction. The fungal spores also were carried by rivers and wind into the oceans.

Fungi are known to be among the first colonizers after forest destruction. So-called fire fungi commonly spread on downed trees after forest fires and other destructive events. For example, soon after the forest devastation caused by the eruption of Mount Saint Helens in 1980, networks of fungi were seen spreading among the fallen logs. Similar widespread growth of fungi on downed trees was observed on the Caribbean island of Montserrat after a volcanic eruption in 1995 destroyed tropical rain forest there. Furthermore, studies have also shown that modern terrestrial fungi proliferate in dying forests that are severely affected by acid precipitation caused by pollution.

In 1998, my fieldwork took me to the Karoo basin in South Africa. The Karoo is a semidesert region known for its stark, beautiful scenery. We were there not to admire the countryside but to try to find the Permian/Triassic boundary in the nonmarine sediments laid down on river floodplains at that time. The rocks were well known and full of reptile fossils that marked the end-Permian extinction. Using geologic maps, we combed the countryside for outcrops that might show the transition. After several days of searching, we decided that a road cut near the small town of Carleton Heights crossed the boundary, with gray-green Permian sediments in a gully just below the roadway and reddish Triassic sediments at and above the road.

Maureen Steiner, an expert on ancient magnetism at the University of Wyoming, and I wanted to determine the magnetic-reversal history across the boundary, so we took a series of oriented core samples for magnetic analysis back in the lab. Unfortunately, when the samples were studied, we found that the rocks had been affected by heat from younger igneous intrusions nearby and did not give accurate magnetic information for the Permian/Triassic boundary interval. I decided, however, to send samples to my colleague Yoram Eshet at the Open University in Jerusalem. He is an expert on ancient pollen and spores and was familiar with the fungal event in his native Israel. We were delighted when, a few months later, we heard that Eshet had found indications of the extinction of almost all forest plants, and the fungal spike marking the extinction boundary, in the Carleton Heights section, just where we had predicted from changes in the sedimentary rocks. The fungal-event zone is about 1 meter (3 feet) thick. At estimated average accumulation rates of about 35 centimeters (14 inches) per 1,000 years for the section, the fungal event would have lasted about 3,000 years in the Karoo.

Studies that investigated ecosystem recovery after the end-Permian extinction report that after the forest ecosystems disappeared, small nonarboreal plants began to spread over the land. Several papers reported a decrease in seed ferns and conifers and an increase in primitive club mosses after the extinction event. The main floral turnover from forests to lowlying vegetation may have taken only a few thousand years or less. On land, the end-Permian extinction correlates with a change in sedimentation, from meandering rivers on wide floodplains to multiple-channel braided streams, choked with sediments, in the earliest Triassic. This change is probably the result of the worldwide loss of vegetation and subsequent increased soil erosion on the continents. In the oceans, the combined effect of a super greenhouse interval (with a temperature rise of 15°C [27°F] in tropical surface ocean waters) and low ocean oxygenation could have contributed to the marine end-Permian extinction, eliminating most primary producers in the oceans.

In the 1990s, I searched for a crater from an impact that may be related to the end-Permian extinction event. I studied maps of Earth's gravity field (the measure of the pull of Earth's gravity), which could show a buried crater (a depression of the dense bedrock, filled in by lighter sediment) as an area of decreased gravity. Such is the situation of the Chicxulub crater in Mexico (see figure 3.9). I was looking for a circular negative gravity anomaly, and I assumed that since the extinctions at the Permian/Triassic boundary were more severe than those at the end of the Cretaceous, the potential crater could be larger than Chicxulub, perhaps 250 to 300 kilometers (155 to 185 miles) in diameter. In such a large comet impact, much of the ejected material may have escaped from Earth, perhaps leaving little in the way of an iridium-rich boundary clay layer.

Published maps of the gravity field over the oceans have been constructed from satellite measurements. In my search, I was struck by a very interesting circular pattern, about 250 kilometers in diameter, on the continental Falkland Plateau, just to the west of the Falkland Islands. The gravity low seemed outline a large basin. This West Falkland Basin shows up in the gravity data as a distinct negative anomaly surrounded by a ring-shaped positive anomaly, both in clear contrast to the local average values on the plateau. Seismic reflection studies of the area, carried out for petroleum exploration, revealed that the structure has the form of a roughly circular basin about 3 kilometers (nearly 2 miles) deep and about 250 to 300 kilometers (155 to 185 miles) in diameter. Mesozoic and Cenozoic sediments fill the basin, so it may be late Paleozoic in age.

The West Falkland gravity anomaly is very similar to that of the Chicxulub impact structure, which was actually discovered from its associated negative gravity anomaly. Like Chicxulub, the proposed Falklands structure has no expression in the topographical maps of the area; no submarine depression is visible. Recently, Max Rocca of the Planetary Society noticed that aeromagnetic regional maps in the same area exhibit an impressive rose-shaped positive magnetic anomaly, about 250 kilometers (155 miles) in diameter, of a kind seen at some large impact structures. At the present time, however, there is no unambiguous evidence for a large impact at the time of the end-Permian event. Instead, evidence points to another potentially cataclysmic cause of mass extinctions: enormous outpourings of lava called flood basalt eruptions.

9

• • • •

Catastrophic Volcanic Eruptions
and Extinctions

Surely, if the CFCs [chlorofluorocarbons] in spray cans can destroy
the ozone layer, the volatiles (including chlorine, fluorine, carbon
dioxide, and sulfur dioxide) from this enormous eruptive event
must have had a significant effect. Can it be just coincidence that
the other major Phanerozoic eruption of the same magnitude
occurred in Siberia at the transition between the Paleozoic and
Mesozoic eras?

CHARLES L. DRAKE AND YVONNE HERMAN, "DID THE DINOSAURS
DIE OR EVOLVE INTO RED HERRINGS?"

The Earth-shaking discovery by Walter Alvarez and his team drew the
attention of geologists in varied fields. Many of these scientists disagreed
with the Alvarez group's conclusions, and the specter of Charles Lyell was
raised against anyone who would go so far as to present a geological hypothesis that involved an extraterrestrial catastrophe. Critics argued that the
evidence for impact was not convincing because it relied on precise measurements of the exotic element iridium, which most geologists knew nothing about. Early and vocal criticism of the Alvarez hypothesis came from
geologists Charles Officer and Charles Drake at Dartmouth College. In articles published in *Science* in 1983 and 1985, they argued that the supposed
impact layer at the Cretaceous/Paleogene boundary did not represent a
worldwide, instantaneous occurrence, and they claimed that the extinctions were better explained by widespread volcanism at the end of the
Cretaceous. This became known as the "volcanist" position.

I was teaching Earth science at Dartmouth when news of Alvarez's
hypothesis broke in 1980, and I remember sitting in Drake's office while
he lamented the impact idea and the intrusion of physicists like Luis

Alvarez into a geological question. Drake was against the impact explanation right from the start, and he began searching the geological literature to find any evidence that could be interpreted as refuting the impact idea. Again, our uniformitarian geological bias, dating back to Lyell, made catastrophic and extraterrestrial explanations for geologic events an almost visceral repulsion for some geologists.

To get an idea of the opposition to the impact hypothesis, it may be pertinent to look at how geologists treated another revolutionary idea, continental drift. In 1949, Traugott Wilhelm Gevers (in a tribute to Alexander du Toit, an early pioneer of drift) wrote:

> In certain quarters the Theory of Continental Drift has come to be regarded as being in the nature of a monstrosity inducing abhorrence and nausea, even of a moral turpitude unmentionable in "nice" geological society. Adherence to the new heresy immediately put one outside the pale of rational human beings, to say nothing of refined intellect. Any attempt to show that "there might be something in it" was treated with scorn and derision and led to reflections on the adequacy of one's decent bourgeois upbringing.

In the early 1980s, the response in some quarters to the idea that extraterrestrial impacts were involved in mass extinctions was similar.

In the view of Officer and Drake, the Alvarez group had not done its geology homework. The boundary layer, they claimed, was derived from volcanic eruptions, which had released the iridium and had even produced the shocked quartz. They noted that high levels of iridium had been discovered in vapors coming from volcanoes in Hawaii and from the volcanic island of Réunion in the Indian Ocean. These are hotspot volcanoes, presumably drawing magma from deep in Earth's interior, and the Réunion hotspot had been the source of the Deccan Traps flood basalt eruption near the time of the Cretaceous/Paleogene boundary in India (figure 9.1). ("Trap" comes from *treppen*, the Swedish word for "steps," as the outcrops of the congealed lava flows in flood basalt provinces often occur as stair steps in the topography; figure 9.2.) The huge Deccan eruptions produced more than 1 million cubic kilometers (240,000 cubic miles) of magma. Imagine a field of lava flows as extensive as the state of Texas, erupting from gaping fissures and building up a plateau of multiple lava flows more than 1,000 meters (3,280 feet) thick (figure 9.3). It is reasonable to think that the environmental effects of such giant eruptions could be severe.

A very outspoken critic of the impact idea, Officer claimed in several papers that large explosive eruptions, or super-eruptions, such as the Toba event in Indonesia (about 74,000 years ago) could produce shocked quartz. It turns out that the volcanic ash layer in the Indian Ocean, generated by the Toba eruption, contained quartz grains with only a single set of poorly developed deformation features rather than the multiple, well-developed sets found in quartz shocked by hypervelocity impacts (see figure 3.4). Also, although Toba was a very large explosive eruption, it caused no extinctions. It is telling that the volcanist position was never widely accepted

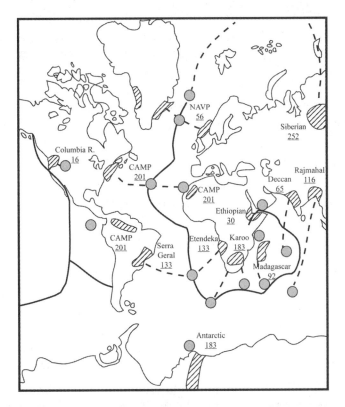

FIGURE 9.1 The distribution of continental flood basalts of the past 250 million years, and related hotspots.

FIGURE 9.2 Iguazu Falls, on the border between Brazil and Argentina. The stair steps in the falls are the eroded individual flows of the Serra Geral flood basalts (133 million years ago).

by working volcanologists, who know something about the pressures involved in volcanic eruptions. At most, eruptions reach only a few tens of bars of pressure (one bar being equal to atmospheric pressure), whereas impacts create tens to hundreds of kilobars of pressure—the intensity needed to produce true shocked minerals. Furthermore, volcanic eruptions are depressurization events, not hypervelocity shock events. The bottom line is that calculations show that volcanism could not have produced the Cretaceous/Paleogene boundary layer's large amounts of iridium and other trace elements (in cosmic proportions) or its shocked minerals.

Despite the evidence for impact, several people continue to maintain that the Cretaceous/Paleogene extinctions were primarily related to the environmental effects of the volcanic Deccan Traps. For example, Dewey McLean, a paleontologist at Virginia Tech, was one of the first to suggest a connection between the paroxysmal Deccan eruptions and the end-Cretaceous extinction boundary. At a conference in 1979, even before the impact layer was discovered, he proposed that a global warming from increased atmospheric carbon dioxide was to blame for the extinctions. In a later

FIGURE 9.3 A section through the Deccan Traps in India at Maha-baleshwar, showing the multiple lava flows, tens of meters thick, making up a stack more than 1,000 meters (3,280 feet) thick.

paper, he noted the close occurrence of the extinctions with the Deccan lava flows and proposed that the eruptions released large amounts of carbon dioxide. In McLean's view, there was no need for the impact hypothesis, so he ignored the evidence for impact in the latest Cretaceous.

Could flood basalt eruptions lead to mass extinctions? In 1988, Richard Stothers of NASA and I published an article in *Science* showing that some continental flood basalts seemed to be correlated with extinction events over the past 260 million years (within the error margins in dating the lava flows and extinction events at the time). As it turns out, massive volcanic lava flow eruptions and their environmental effects *are* potential candidates for causing some of the mass-extinction events. In recent years, as methods of radiometric dating have improved, and as the flood basalts have been more closely studied, the correlation between basalt eruptions and some extinction events looks even better (table 9.1).

These massive eruptions are linked to the rise of great plumes of heated, semisolid rock in Earth's mantle, which create volcanic hotspots on Earth's surface and fracture the crust. Flood basalts can apparently erupt through radiating dike swarms related to subsequent continental rifting. The Deccan Traps can be interpreted in this way (figure 9.4). In India, the focal point of the radial pattern is Mount Girnar, a circular, layered igneous intrusion of unknown origin. It seems as though Earth's crust was fracturing, opening up large fissures that fed the widespread lava flows. Similar associations of major volcanic lava flow eruptions with radiating dike swarms can be found elsewhere on Earth, as well as on Venus and Mars. Hotspot initiation is connected with a number of geologic events in addition to the flood basalts, such as continental rifting, changes in

TABLE 9.1 Ages of Continental Flood Basalts over the Past 250 Million Years, and Correlation with Extinctions and Anoxic Events

Flood Basalt	Age (millions of years ago)*	Extinction or Anoxic Event
Columbia River	16	
Ethiopian	30 ± 1	
North Atlantic	56	Anoxic?
Deccan	65.5	Extinction, anoxic?
Madagascar	92 ± 1	Extinction, anoxic
Rajmahal	116 ± 1	Extinction, anoxic
Serra Geral/Etendeka	133 ± 1	Extinction?, anoxic
Karoo-Ferrar	183 ± 1	Extinction, anoxic
Central Atlantic Magmatic Province	201 ± 1	Extinction, anoxic
Siberian	252	Extinction, anoxic
Emeishan	260	Extinction, anoxic

*Ages after compilation by the author.

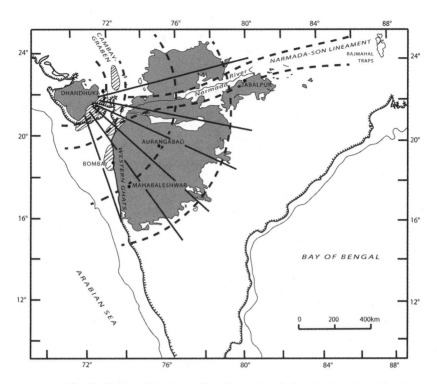

FIGURE 9.4 The distribution of the Deccan flood basalts in India (*gray*). The origin of the flows through a radial dike swarm is proposed.

seafloor-spreading rates, tectonics, and sea-level variations. These occurrences tend to be correlated, and any one of them may have contributed to extinctions of life.

The geologic record shows 11 continental flood basalt provinces over the past 260 million years (an average of about one every 24 million years). Using the results of increasingly more accurate radiometric dating methods, and studies of the magnetism of the lava flows, volcanologists have shown that the bulk of the flood basalt episodes erupted over quite brief periods of geologic time—less than a few hundred thousand years in most cases, and perhaps considerably less for the most voluminous phases of the eruptions.

In a recent study combining evidence for extinctions, impacts, and continental flood basalts, Ken Caldeira and I compared the timing of 13 extinction events over the past 260 million years with the ages of flood

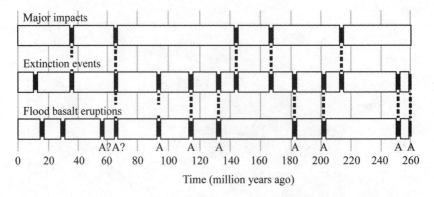

FIGURE 9.5 Correlation among large impacts, flood basalt eruptions, and extinction events. All six of the largest impacts of the past 260 million years correlate with extinctions, and eight out of 11 flood basalt eruptions correlate with extinctions. Seven ocean anoxic events (*A*) are correlated with flood basalts and extinctions.

basalt eruptions and the largest impact craters. We found that eight of the extinction events were correlated with the dates of the largest flood basalt eruptions, whereas the six largest impact craters are coincident with other times of extinction (figure 9.5). The odds of these correlations happening by chance are astonishingly low (less than 1 in 100,000). All in all, large impacts and flood basalt eruptions could explain 12 of the 13 extinction events of the past 260 million years. It seems that extinctions may have been triggered by catastrophic events both from above and from below. In one case, the Cretaceous/Paleogene boundary, there is a coincidence of both a very large impact (Chicxulub) and a large flood basalt eruption (Deccan Traps), suggesting a possible one-two punch to the biosphere.

Furthermore, close correlations exist between the ages of some flood basalts and the times when large regions of the oceans became depleted of dissolved oxygen—so-called oceanic anoxic events. One idea is that global warming produced by carbon dioxide emitted by the flood basalt eruption led to sluggish oceans that became stagnant and low in dissolved oxygen. These times include the Late Cretaceous anoxic event at about 94 million years ago, roughly correlated with the eruption of the Madagascar flood basalts; an Early Cretaceous anoxic event at 113 million years ago, roughly correlated with the Rajmahal basalts in India; a Middle Jurassic anoxic event at about 183 million years ago, correlated with the Karoo-Ferrar basalts in

South Africa and Antarctica; an anoxic event at the Triassic/Jurassic boundary, 201 million years ago, correlated with the Central Atlantic Magmatic Province; and the widespread end-Permian "superanoxic event" at about 252 million years ago, which coincides with the eruption of the Siberian basalts (see figure 9.5). Anoxia in surface ocean waters could have been deadly for planktonic organisms and those living in shallow water. The release of toxic hydrogen sulfide by green sulfur bacteria in the anoxic waters may have been another factor.

With regard to the massive Siberian Traps from the end of the Permian period, paleontologists interpret the fossil evidence as placing the time of the end-Permian mass extinction close to the time of the lowest lava flows. The most reliable age determinations for the basalts agree closely with radiometric ages of volcanic ash layers associated with the latest Permian extinctions, found in the end-Permian geologic type section in China (see figure 8.4). Geochronologists Seth Burgess and Sam Bowring at MIT, using new precise age determinations, recently estimated that the eruptions lasted about 300,000 years, starting before the mass-extinction event and then continuing after the extinctions.

Is there a place for volcanism in the Cretaceous/Paleogene story? The Deccan eruptions seem to have occurred in several major phases. Recent paleomagnetic results and careful radiometric dating suggest that the largest flows were extruded rapidly, close to the time of the Cretaceous/Paleogene boundary. Geochemist Gregory Ravizza and his colleagues at the University of Hawaii documented, in an ocean core, an increase in volcanically derived osmium just prior to the Cretaceous/Paleogene boundary. The osmium provides a proxy for the Deccan volcanism. Detailed climate data from the same core suggest that the environmental consequences of the Deccan Traps included a transient global warming of 3°C to 5°C (5°F to 9°F), which occurred prior to and separate from the end-Cretaceous impact event, though this was probably not enough, on its own, to trigger a major mass extinction of life.

The eruptive style of flood basalts is important, when considering the atmospheric and climatic effects. Originally, flood basalt lavas were envisioned to flow as turbulent sheets, up to 100 meters (328 feet) thick, that covered large areas in a matter of days. However, in the 1990s, Stephen Self, a volcanologist now at Berkeley, along with his colleagues, found that flood basalts erupt mainly as fluid pahoehoe-type flows that form a surficial cooling crust and then actually "inflate" with lava to their greatest thicknesses (meters to tens of meters) as they reach their greatest extent. Self is one of the best field geologists working on volcanic rocks. His findings suggest eruptions of individual flows over periods of time ranging from years to decades, or perhaps longer. For example, based on the peak output rates of a large historical lava flow eruption, the Laki event in Iceland in 1783, the time required for eruption of one large flood basalt lava flow would be in the range of tens to hundreds of years. What's more, the Laki basalts were erupted from lava fire fountains, where the magma could release its gases as it jetted into the air. In flood basalt eruptions, very high fire fountains would allow those volatiles to reach high into the atmosphere.

I have visited seven flood basalt provinces on five continents, and they are all quite similar, comprising multiple lava flows of fluid basaltic composition (see figure 9.3). Some are found today on adjacent continents, now separated by ocean, where the volcanism occurred just prior to the splitting of the continental masses. I once helped create a cinematic dramatization of this process. During the filming of a documentary about the opening of the South Atlantic by the unzipping of the southern continents, the director had me sample the Serra Geral basalts in Brazil, located in a tropical rain forest, and then walk out of the shot, carrying my sample. Six months later, in the desert of Namibia, I walk into the shot (presumably after traversing the Atlantic Ocean!) and place my piece of basalt next to an outcrop of the coeval Etendeka lava flows. The rocks are the same in kind and age but are now separated by the South Atlantic. In the middle of the ocean lies a volcanic island, Tristan da Cunha, the present location of the hotspot that caused the eruptions about 133 million years ago.

It has been suggested that flood basalt eruptions may produce several kinds of direct environmental effects, beyond the greenhouse warming, resulting from volcanic emissions of carbon dioxide. For instance, a significant amount of methane, an even more potent greenhouse gas, can come from interactions between the flood basalt magma and organic-rich sedimentary deposits, such as coal or oil shale, which seems to be a not-uncommon occurrence during intrusions of basaltic magma. Henrik Svensen and his colleagues at the University of Oslo found numerous hydrothermal vents from the explosive release of gases associated with the eruption of the North Atlantic basalts, the Karoo basalts, and the Siberian Traps. A substance similar to coal fly ash, produced by coal combustion, has been detected in some Permian/Triassic boundary sections. The volcanically induced greenhouse warming could have led to the secondary release of additional methane from unstable methane hydrates stored in continental shelf sediments, exacerbating the warming.

A novel idea about the end-Permian extinctions was published in 2014 by Daniel Rothman from MIT and his colleagues. Trace-metal analysis of latest Permian sedimentary rocks in several regions turned up anomalous concentrations of nickel. In Austria, two nickel-abundance anomalies correlate with two negative shifts in carbon isotopes in the oceans (see figure 8.2), suggesting a potential relationship between nickel-rich emissions from the Siberian Traps and carbon cycle disturbances. My colleague Sedelia Rodriguez of Barnard College and I have traced the end-Permian nickel anomaly to other Permian/Triassic sections, confirming the global distribution of the nickel peak (or peaks). The Rothman group proposed that the nickel enrichment in the oceans may have led to proliferation of methane-producing microbes, which are normally limited by available nickel. The dissolved-nickel enrichment in the oceans would have allowed the microbes to spread and to effectively degrade organic material, with the emission of copious amounts of methane enriched with light carbon-12. This methane release could have contributed to the carbon isotope shifts seen at the end of the Permian period.

The addition of large amounts of carbon dioxide into the ocean–atmosphere system over a relatively short period of time can also cause changes in seawater acidity. We see this happening at the present time as a result of anthropogenic release of carbon dioxide. Noted geochemist Robert Berner at Yale and co-workers used computer simulations to estimate the degree of acidity that may be produced in the surface ocean by the release of carbon dioxide and sulfur dioxide from massive basaltic eruptions. They concluded that the gases from the Central Atlantic eruptions at the end of the Triassic period could have produced an acidic surface ocean that persisted for 20,000 to 40,000 years, creating a problem for organisms that produce shells of calcium carbonate. Other extinction episodes, such as the end-Permian event, are marked by preferential survival of organisms with resilience to ocean acidity but by major extinction of the more vulnerable reef builders.

Another potential effect of flood basalt eruptions is short-term climatic cooling, primarily as a result of the formation and spread of sulfuric acid aerosols in the atmosphere. These small droplets form from sulfur dioxide gas injected into the upper atmosphere by eruption plumes rising above the volcanic vents and fissures. These aerosols have a short residence in the lower atmosphere, but those that reach the upper atmosphere may persist for several years, where they primarily backscatter incoming sunlight, cooling the planet. Could this cause a "volcanic winter"?

In the case of the Siberian Traps, reactions between magma and anhydrite (calcium sulfate) deposits may have added significant sulfur dioxide to the eruption emissions. Studies using a global climate model suggest that, if flood basalt eruptions were continuous at an average activity level for decades, then significant cooling of climate (estimated at 4°C to 5°C [7°F to 9°F] over a period of about 50 years) might be possible. However, with hiatuses of perhaps thousands to tens of thousands of years between individual flood basalt eruptions during the formation of a flood basalt province, even large releases of sulfur gases might not produce a cumulative and lasting effect on climate and the environment.

The Laki eruption caused noticeable haziness and dimming of sunlight in Europe, according to Benjamin Franklin, who was in Paris at the time. This so-called dry fog was seen as far away as China. Climatic cooling followed, and the winter of 1783/1784 was the coldest ever recorded in the eastern United States. The local haze over Iceland contained high concentrations of sulfuric, hydrochloric, and hydrofluoric acids, causing skin lesions in animals and humans, stunted grass, dead trees, and loss of 50 percent of livestock, most likely from fluorine poisoning. The accompanying "haze famine" led to the death of 20 percent of the Icelandic population. Northern Europe also reported acid rain effects, which affected the growth of vegetation.

Scaling up from Laki (with a flow volume of about 10 cubic kilometers [2.4 cubic miles] of magma), it seems clear that acid rain could be a serious problem during flood basalt episodes with volumes of 100 to 1,000 cubic kilometers (24 to 240 cubic miles) of lava. As we have seen, the end-Permian extinction is marked by the global die-out of vegetation, and this might have been related to acid rain created by the Siberian Traps eruption.

Thus cataclysmic volcanism in the form of flood basalt eruptions may have produced environmental changes so severe as to have caused or contributed to a number of mass extinctions. The good correlation among flood basalt eruptions, some extinction events, and oceanic anoxic events supports this conclusion. However, as we will see, it is possible that both the eruptions and the extinctions may be products of cyclical changes in Earth's movements in the Milky Way galaxy.

10

• • • •

Ancient Glaciers or Impact-Related Deposits?

Accurate identification of ancient pebbly sediments as tillites is
not possible by lithology alone, as such rocks can also be produced
by nonglacial agents.

L .J. G. SCHERMERHORN, "LATE PRECAMBRIAN MIXTITES"

At the present time, Earth is in the grip of an ice age. Just 20,000 years ago, at the time of maximum ice advance, thick glaciers covered the northern continents down to the latitude of New York City. Luckily, we are now in an interglacial period of relatively warm climate, but the ice will be back in a few thousand years. The current 2 million-year-old ice age can be traced back to a cooling of Earth's climate that began about 30 to 35 million years ago. Prior to that, Earth enjoyed a long period of warm climate.If one studies the geologic record, however, one can find deposits ascribed to previous ice ages, as far back as 2.9 billion years. What do ancient glacial deposits have to do with catastrophism?

Glaciation in the present ice age produces typically bouldery and chaotic sediments attributed to glacial erosion and deposition; these deposits are called glacial till. Ancient glacial deposits, called tillites (figure 10.1), are abundant at certain times in Earth's history and are used to define previous ice ages in deep time, when large portions of Earth were supposedly covered by ice. In the 1990s, a few papers were published, including one of my own, suggesting that some deposits in the deep geologic past, now interpreted as sediments produced by glaciers, may in actuality have formed by nonglacial processes involving mass flows called debris flows and debris falls. This includes some debris-flow deposits potentially related to large-body impacts. These papers stirred up a bit of geological controversy at the time.

FIGURE 10.1 The Elliot Lake tillite (Late Precambrian) in southern Canada, a probable debris-flow deposit.

Debris flows are sediment flows driven by gravity, in which the larger grains (up to the size of a very large boulder) are carried along by a thick slurry of water and fine sediment. Debris-flow deposits commonly display distinct features, such as alignment of stones, that give evidence of rapid flow. They occur along with rhythmically bedded siltstones and mudstones, commonly containing outsize stones provided by debris falls, where stones have moved by gravity down slopes and/or where the large stones outran the debris flows.

Cratering expert Verne Oberbeck and his colleagues at NASA calculated that a significant volume of debris-flow deposits created by past impacts, like the Albion Formation at the Cretaceous/Paleogene boundary and the Bunte Breccia from the Ries impact in the Miocene epoch, should have accumulated from the many impacts on Earth over geologic time. They noted, however, that only a few relatively recent impact-related debris-flow deposits have been identified, whereas ancient glacial tillites are abundant in the geologic record. Where are the expected impact deposits? Could some tillites be misidentified impact ejecta deposits?

Over the past 40 years, tillites of various ages, from Archean to Late Paleozoic, have been reinterpreted as submarine debris-flow deposits, formed at the glaciers' terminus, with rainout of ice-rafted debris and potential tectonic influences supplying much of the coarse debris that was processed by the glaciers. Since some tillites have been interpreted as debris-flow deposits, we might ask: What are the criteria for determining a glacial origin for these deposits? The classical answer is to point to striated or scratched pavements beneath the tillites, presumably due to glacial erosion; to faceted, polished, and striated stones; and to outsize stones in fine-grained deposits, commonly interpreted as stones dropped from glacial icebergs. Not all recognized tillites have all these features, but most have some.

There is now evidence, however, that nonglacial debris flows can produce polished, faceted, and striated stones as well as striated pavements and sculpted features on bedrock. A potential example is the famous Bigganjargga tillite at Varangerfjord in northern Norway. This tillite, more than 700 million years old and resting on what appears to be a striated pavement, has been reinterpreted by some as a debris-flow or mudflow deposit that created the striations on the underlying bedrock pavement, with no direct evidence for a glacial origin.

The feature probably considered most indicative of glaciation is the presence of outsize stones in finer-grained sediments, interpreted as dropstones from icebergs. We now know, however, from the work of S. B. Kim and colleagues at Seoul National University, that debris flows and related debris falls can deposit outsize stones that mimic ice-rafted debris. They studied the Kyokpori Formation (Late Cretaceous) in southwestern Korea, which consists largely of sandstones and siltstones deposited in an ancient lake basin in a delta environment. The lowermost part of the formation is dominated by bouldery deposits and sandstones with outsize stones. The outsize stones are randomly scattered within the sandstone beds, occasionally forming clusters. Furthermore, laminated deposits of alternating silt and mud with larger stones, commonly ascribed in tillite sequences to quiet deposition in proglacial lakes or the sea and dropstones from floating ice, might instead be formed by deposition from fast-moving, intermittent muddy currents and debris falls.

Known coarse impact ejecta deposits also show features of debris-flow and debris-fall deposition, and they have many of the characteristics of rapid mass flow. In fact, several deposits that were initially ascribed to glaciation have been reinterpreted as the products of impacts. For example, the 15 million-year-old Bunte Breccia is now known to be a product of debris flows related to the Ries impact structure in southern Germany. For decades, this deposit was thought to be the result of glaciation. The base of the Bunte Breccia contains very large, highly fractured limestone blocks, and the underlying bedrock is cracked and fissured. Most striking, in places where the breccia was deposited on competent limestone rock surfaces, the surfaces are scoured and sculpted and show linear striations generally pointing radially away from the Ries crater.

Furthermore, pebbles from the Bunte Breccia show striations and mirrorlike polish, once thought to be strictly glacial in origin. I have seen similar polished and striated pebbles of limestone in Belize, in the ejected material of the Chicxulub crater. In another case, coarse deposits of Mesozoic age, covering up to 6,000 square kilometers (2,317 square miles) in Russia, were initially classified as glacial debris but are now known to be impact deposits associated with the 80-kilometer-diameter (50-mile) Puchezh-Katunki impact structure, dated at about 167 million years ago (in the Middle Jurassic).

Most significantly, Oberbeck and his colleagues made a cursory sampling of the Dutch Peak tillite in southern Utah (attributed to Late Precambrian glaciation, about 700 million years ago) and discovered a stone clearly exhibiting shock deformation features in quartz. Such shocked stones are rare, even in known impact deposits such as the Bunte Breccia, so this discovery should have raised some doubts about a glacial origin for the Dutch Peak deposits.

How can we discriminate between debris flows of impact origin and debris flows from other causes, including glaciation? The best evidence would be shocked minerals in the deposits, but the shocked stones may be very rare and could be reworked by erosion of older impact structures. I propose that another potential indicator of impact origin of debris-flow sediments is the presence of stones showing brittle failure after initial plastic deformation under extreme stress. These stones exhibit deformed shapes and have parallel displaced and rotated fractures, often cutting through only part of the stone so that the stone remains hinged but intact (commonly called a "bread-cut-to-slices" structure). Deformed stones also may be entirely crushed and/or broken apart. These deformations are not a result of normal tectonism—they occur in deposits that have not been tectonically deformed.

I have sampled such hinged, fractured, and crushed stones from the Bunte Breccia (figure 10.2); from the Late Permian Araguainha impact structure in Brazil, where the pebbles show features of hypervelocity shock (figure 10.3); and from the Albion Formation in Belize, associated with the Chicxulub impact structure. These distorted and broken stones seem to require deformation under temporary, very high confining pressures, estimated by geochemist Ed Chao of the U.S. Geological Survey at up to 30 kilobars, which can be reached by hypervelocity impacts. These features could not form in ordinary glacial deposits.

A search of the literature on ancient glacial deposits revealed several potential candidates for reclassification of a supposed tillite as an impact deposit, based on the presence of deformed, fractured, and hinged stones. Two examples stand out, as they are Middle Proterozoic in age (1 to 1.5 billion years old), which is not a time of accepted widespread glaciation. In fact, the acceptance of these deposits as glacial would mean a serious reevaluation of Earth's climate history. The approximately 50-meter-thick (164-foot) Gangau tillite in north-central India (1.2 to 1.4 billion years old) contains pebbles of chert and quartz that reportedly show brittle failure after deformation. One report showed deformed boulders, with rotated en echelon, offset fractures ("bread-cut-to-slices" structure), from somewhat earlier Middle Proterozoic deposits in India (figure 10.4), where the stones were apparently being squeezed and then partly failed in a brittle manner.

Another potential Middle Proterozoic impact debris-flow deposit is the Stoer Group in Scotland (1.2 billion years old), which has severely fractured bedrock at the base of debris-flow deposits containing stones with rotated fractures and is attributed to glaciation. The Stac Fada Member within the Stoer Group, however, has already been recognized as melted material ejected from an impact. A potential crater occurs nearby. The supposed tillites are quite likely to be ballistic debris flows related to that impact.

FIGURE 10.2 Jurassic limestone cobble from the Bunte Breccia (Ries crater, Germany), showing hinged fractures and incipient crushing.

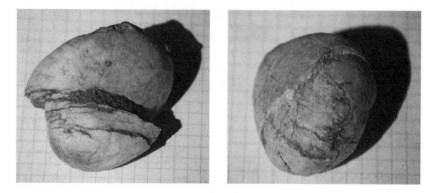

FIGURE 10.3 Pebbles from the Araguainha impact structure in Brazil, showing offset fracturing after plastic deformation. These pebbles show evidence of hypervelocity shock (feather features in quartz).

The Luoquan tillite of central China (about 530 million years old, from the Early Cambrian, also not a time of accepted glaciation) displays stones with similar displaced parallel fissures. These are referred to as "cracked gravels." The somewhat older, 60-meter-thick (197-foot) Nantuo tillite contains fractured and hinged stones, described as marked by striations, pressure pits, pressure crevices, polished surfaces, and concave fractures, similar to features seen on stones in the Chicxulub Albion deposits. Some of

FIGURE 10.4 "Bread-cut-to-slices" structure in a stone from a Middle Proterozoic tillite in India. (From F. Ahmad, An ancient tillite in central India, *Quarterly Journal of the Geological, Mining and Metallurgical Society of India* 27 [1955]: 157–61)

the stones in the Nantuo tillite are described as crushed. Several other studies of ancient glacial deposits describe tillites that show deformed, fractured, and hinged stones, or describe stones as "broken, fluidized, squeezed, and deformed."

The recognition that many tillites in the geologic record have a debris-flow origin, and problems with criteria previously used for glacial origin, suggest that a reinterpretation of some supposedly glacial tillites as nonglacial deposits may be necessary. Debris flows from known impact-cratering events contain characteristic deformed, en echelon fractured, hinged, and crushed stones produced by deformation followed by brittle failure, which seems to require high, temporary confining pressures. The presence of such deformed and fractured stones in some tillites could be a potential indicator of an impact origin for those deposits. So it seems inescapable that some ancient debris-flow deposits will prove to have been produced by large-body impacts. A careful study of proposed tillites and associated deposits, to search for deformed and fractured stones as well as evidence of shock deformation, seems warranted.

11

• • • •

The Shiva Hypothesis

COMET SHOWERS AND THE GALACTIC CAROUSEL

Our conclusion is that there have been pulses in the impact rate
on Earth, some of which are correlated with mass extinctions, and
some probably are due to comet showers.

EUGENE SHOEMAKER AND RUTH WOLFE, "MASS EXTINCTIONS,
CRATER AGES, AND COMET SHOWERS"

If mass extinctions are periodic, as David Raup and Jack Sepkoski claimed, then the major question is what might be causing the periodicity. In September 1983, astrophysicist Richard Stothers and I read with great interest the early reports of the mass-extinction results in several news magazines. After Raup sent us a prepublication copy of their paper on extinction periodicity, Stothers and I began searching for possible causes for periodic mass extinctions.

Stothers was a polymath, an astrophysicist who also read ancient literature in the original Latin and Greek to find records of changes in the atmosphere, sunspot observations, volcanic eruptions, and other natural events in classical antiquity. We had already collaborated on a compilation of accounts of the effects of volcanic eruptions on the atmosphere and climate in ancient times. Stothers discovered the great "dry fog" and dimming of the sun that occurred in the cold year 536 C.E., resulting from one of the largest volcanic eruptions in historical times. Some writers even suggested that the severe atmospheric perturbation in that year was caused by an impact.

Geologic cycles had long drawn my attention. I had in my office a file folder full of papers on the subject. Many articles propose long-term geologic cycles of various kinds and of various lengths, but most of these ideas have been attributed to the imagination of avid geologists seeing patterns where none exist. Few of these purported cycles survived rigorous statistical analysis. Cycle chasing is a risky game, but Raup and Sepkoski's research seemed solid and cried out for a solution.

If the extinctions are periodic, and impact caused the end-Cretaceous event, then could catastrophic impacts themselves be periodic? By late 1983, only two extinction events (the end-Cretaceous and the Late Eocene) were known to be associated with impacts. We knew, of course, that iridium anomalies, shocked quartz, and microtektites were not the only evidence for collisions of large bodies with our planet; a number of impact craters of various sizes and ages mark the location of past impacts (figure 11.1), and the dates of several seemed to coincide fairly well with some of the mass extinctions. For example, as Harold Urey noted in 1973, the large 100-kilometer-diameter (62-mile) Popigai crater in Siberia, which created shocked quartz and microspherules, was known to date from about 36 million years ago, close to the time of the Late Eocene extinction event.

Stothers and I dug out the most complete list of terrestrial impact craters, compiled by cratering specialist Richard Grieve of the Geological Survey of Canada (figure 11.2). In 1983, the list included about 150 documented impact craters, their sizes, their locations, and an estimate of their ages. Today, the ever-growing list is maintained online as the Earth Impact Database and includes about 185 craters. Even so, these craters represent only a small subset of the actual number of bolides that have collided with Earth. Many more impact craters are so severely eroded and/or covered by sediments that they are difficult to identify. What's more, no craters have been found in the deep ocean, only in shallow continental shelf areas. This is not surprising, because the ocean floor is young, at most only 185 million years old, and thus should exhibit relatively few craters. In addition, no one knows how an impact into thin ocean crust might be expressed. I once argued that a large impact might penetrate the ocean crust and be followed by a period of mantle-derived volcanism that could mask the original crater.

Many of the estimates of the age of the craters are merely rough limits based on the age of the rocks targeted by the impact or the age of the oldest sediments burying the impact structure. But we decided that a number of Grieve's craters were dated well enough, through radiometric age determinations, for use in a rigorous statistical analysis of the timing of

(a)

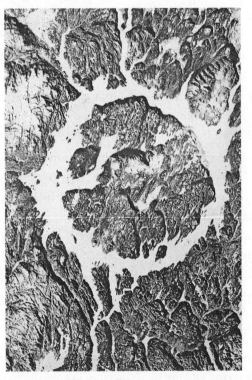

(b)

FIGURE 11.1 Impact craters: (a) the small Meteor Crater in Arizona, which is 1.2 kilometers (0.75 mile) in diameter and about 50,000 years old; (b) the large Manicouagan Crater in Quebec, which is 100 kilometers (62 miles) across and about 214 million years old.

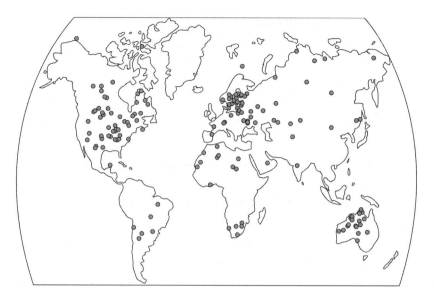

FIGURE 11.2 The distribution of recognized impact craters in the world. (Redrawn from the Earth Impact Database)

the impacts. Stothers had invented a statistical method that he used to look for sunspot cycles in ancient times and that also can be used to search for periodicities in a noisy and incomplete record like that of impacts. (By a strange coincidence, the same method was developed independently by Raup and was used in his and Sepkoski's groundbreaking analysis of mass extinctions.) We decided to run the ages of the best-dated craters in Grieve's list through the NASA computer, using this new method of spectral analysis. When Stothers returned from the computer room with the output of our initial crater analysis, his excited manner and the smile on his face told me that we had struck gold—the impact crater record showed a strong periodicity of about 31 million years.

More recently, in 2005, Korean scientists Heon-Young Chang and Hong-Kyu Moon, using a different technique of analysis, found a 26 million–year period in craters, and Ken Caldeira and I detected a similar cycle, in 2015. Of course, these results are still controversial. Furthermore, impacts may come in discrete episodes. I compared the data on impact-cratering rates over the past 260 million years with the times of the largest impact craters, and with extinction events, and found an excellent correlation among the three—much too good to be the product of chance (figure 11.3). There seem to be times when impact-cratering rates are high, perhaps indicating comet or asteroid showers, and when large impacts tend to occur. These times also are correlative with extinction events.

Stothers's statistical analysis of the crater ages convinced us that the impacts in our data set were periodic, but where were they coming from? There were two possibilities: Earth-crossing asteroids originally from the asteroid belt between the orbits of Mars and Jupiter, or comets, icy bodies from the distant Oort cloud, which surrounds the sun. We doubted that swarms of asteroids from of the stable asteroid belt could have pelted Earth with regularity. That left the Oort cloud comets, which number in the trillions. Stothers told me about the calculations done by astronomer Jack Hills of Los Alamos National Laboratory, showing that the gravitational perturbations caused by a passing star could shake up the loosely bound Oort cloud comets. This would cause large numbers of comets to fall into the inner solar system—a comet shower—where some could strike Earth. Hills had even suggested that such a comet shower could have caused the demise of the dinosaurs.

But if comet showers were the culprits, why would they show a cycle of about 30 million years? I knew of one cosmic cycle—the revolution of the sun around the Milky Way galaxy—with a period of around 250 million

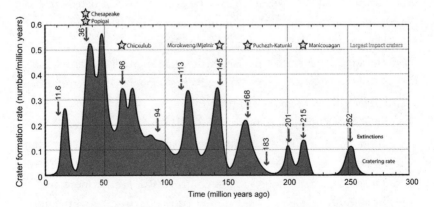

FIGURE 11.3 The history of impact cratering over the past 260 million years. Peaks of impact-cratering rate correlate with mass extinctions (*arrows*) and with the largest craters (*stars*). Minor extinctions are represented by dashed arrows.

years. There might be a shorter cycle related to the sun's passage through the spiral arms of the galaxy, as suggested by British astronomers Victor Clube and Bill Napier, but that period is not well known.

Serendipity often accompanies scientific breakthroughs. It was truly serendipitous that Stothers, an astrophysicist, worked just down the hall from my office at the Goddard Institute for Space Studies. I asked him if there were any known cosmic cycles with a period of about 30 million years. After a few moments of thought, his reply was, "Yes. The motion of the solar system with respect to the midplane of the disk-shaped Milky Way galaxy has a full period of about 60 to 70 million years, and therefore a half-period of about 30 to 35 million years, from one plane crossing to the next. Imagine the solar system as a horse on a carousel." Thus "as we go around the disk-shaped galaxy, we bob up and down through the disk. We pass through the densest part of the disk roughly every 30 to 35 million years" (figure 11.4).

Stothers had once worked on this so-called z-oscillation of the sun and other stars. In my search of the literature, I came across a paper by astronomer Kimmo Innanen at York University in Toronto and his colleagues that gave 33 million years as the best period between plane crossings. Stothers estimated that this value was probably good to plus or minus 10 percent.

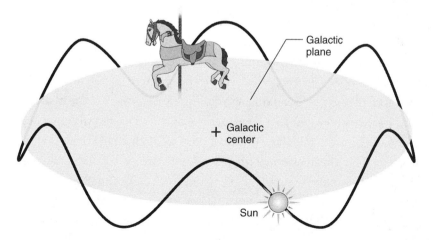

FIGURE 11.4 The carousel-like movements of the solar system around the galaxy and up and down through the galactic midplane.

Considering possible errors in dating extinctions and craters, and uncertainties about the galactic period, the three cycles seem to agree. Surely, it seems too much of a coincidence that the cycle found in mass extinctions and impact craters should turn out to be one of the fundamental periods of the galaxy.

We briefly considered other possible sources of astronomical periodicity, such as solar companion stars or orbiting planets that come close to the Oort cloud, but we rejected these as too unlikely and ad hoc. The connection with the galaxy convinced us that there must be some relationship between these apparent comet showers and the solar system's swings through the galactic plane. It almost seemed too pretty an idea to be wrong. But people searching for cycles have been fooled before, and we still had to answer a question: How does this cycle of movement lead to perturbations of the Oort cloud comets?

We inferred that something must be disturbing the Oort cloud comets at the edge of the solar system and that this object or objects must be very massive to cause such a strong gravitational perturbation. Hills had suggested that a star would do the trick, but close encounters with stars should not take place as often as once every 30 million years. Interstellar clouds of gas and dust are more massive than stars—some more than 10,000 times the mass of the sun. A close encounter with a large cloud might deliver a comet shower.

Our galaxy has a large fraction of its normal, "baryonic" matter arranged in the shape of a flattened disk (figure 11.5). This disk comes about because normal matter cools by emitting photons that carry energy away from the galactic disk. This lowers the velocity of the ordinary matter, and the less-energetic masses move in a smaller volume of space, closer to the midplane of the galaxy. We knew that large interstellar clouds were distributed preferentially near the midplane, so encounters would tend to happen during disk crossing. This tendency provided a key to the formulation of a galactic model to explain the cycles of comet impacts and extinctions.

A comparison of the estimated times when the solar system crossed the galactic plane with the times of impacts and mass extinctions showed a potential correlation. Astronomers Pat Thaddeus of the Goddard Institute and Gary Chanan of Columbia University later argued that the interstellar

FIGURE 11.5 A spiral galaxy, like our own Milky Way, as seen edge-on. Note the concentration of visible matter in the galactic midplane. (Photo by Ken Crawford)

clouds were not concentrated enough in the plane of the galaxy, and the sun's excursions were not far enough away, to create a periodic signal in cratering. But it may be, as some astronomers have suggested, that the sun's excursions from the galactic midplane were greater in the past; most stars of an age similar to that of the sun have considerably greater excursions. The change could have been the result of a past gravitational "kick" to the sun from a near-collision with a giant interstellar cloud.

John Matese at the University of Southwestern Louisiana and his colleagues, however, used computer simulations of galactic motion to calculate that the Oort cloud would be especially vulnerable to gravitational perturbations caused by the galactic tides—in essence, the pull of gravity from all the mass concentrated in the galactic midplane. This was a simpler idea, and it eliminated the need for encounters with interstellar clouds (which also could have occurred, but perhaps more episodically).

More recently, in 2014, Lisa Randall and Matthew Reece, astrophysicists at Harvard University, concluded that the largest gravitational perturbations of the Oort cloud could be from an invisible thin disk of exotic dark matter. Dark matter is a form of matter that astronomers believe may account for about 85 percent of all the matter in the universe. Amazingly, all the visible matter in stars, nebulae, and galaxies accounts for only 15 percent of the matter.

The evidence for dark matter in the universe comes mostly from the motions of galaxies. Dark matter explains the fact that stars far from the center of rotating galaxies have much higher velocities than predicted from the distribution of visible matter alone. Without some additional matter, the galaxies would fly apart. The distribution of dark matter required to explain the "excess" velocity suggests a spherical concentration of dark matter surrounding the galaxies. Evidence for dark matter also comes from study of galaxy clusters, where more matter than is visible is required to explain the gravitational forces holding the clusters together. Dark matter also makes its presence known through gravitational lensing of distant galaxies, where distortion of light from galaxies far behind the dark matter halo of a nearby galaxy forms a ring of mirages around the closer galaxy.

So astrophysicists are convinced that dark matter exists, and most believe that it is most likely composed of weakly interacting massive particles. It does not interact with electromagnetic radiation, so it is difficult to detect. Although it is inferred to exist in spherical halos surrounding spiral galaxies like our own, Randall and Reece suggested that some dark matter would also be concentrated in a thin disk along the midplane of the galaxy. They propose that a fraction of the disk's dark matter may have interactions similar to that of normal matter, dissipating energy and thereby cooling into a very thin disk embedded within the disk of ordinary matter. It is also predicted, by some, that the dark matter disk will naturally fragment into smaller, denser clumps. Such thin-disk dark matter is invisible, but its presence is detected by its gravitational pull on other objects. A test of the existence of a dark matter disk will rely on new data from the *Gaia* satellite, which is now measuring the positions and motion of stars in the galactic plane. The behavior of these stars is dependent on the total mass in the galaxy's disk, which should tell us how much dark matter is present.

With regard to the creation of comet showers, the idea is that when the solar system passes through the densely populated galactic disk, the concentrated gravitational force of the mass (both dark and visible) jostles the Oort comet cloud. This sends a barrage of comets toward the inner solar system about every 30 million years, some of which eventually hit Earth. Where are we in the cycle today? We are relatively close to the galactic midplane, having just crossed it from "below." It takes a couple of million years for a comet to fall from the Oort cloud into the inner solar system. Our precarious position is in line with the dates of several young craters and microtektite layers from the past 1 or 2 million years.

The history of the publication of our original paper on the galactic oscillation hypothesis, in 1984, is worth telling, as it is a good example of the ups and downs of publishing a paper on a hot topic. Stothers and I wrote our results on galactic comet showers in record time in the fall of 1983, including our statistical analysis supporting a 30 million–year cycle detected in impact craters. We submitted our manuscript to the British journal *Nature* in mid-November of that year. We knew that this was a very hot research subject. With Raup and Sepkoski's results reported in *Science, Science News*, and even the *New York Times* by December, surely other groups must be searching for a mechanism for periodic extinctions.

Publication is a waiting game. Stothers and I rushed to get the paper written and submitted, but peer review takes time. We considered sending out preprints of our manuscript, and perhaps even sending copies to the scientific press, but *Nature*'s acceptance requires that no release of information to the press be made prior to the publication of an article. So we sat tight and waited for the reviews from *Nature*.

The reviews came back in mid-December. They crushed our hopes of quick publication. One reviewer, who was brutally critical of our results, suggested that publication be granted, but only after major revision. He thought our time series analyses were "very naive" and our ideas of periodicity of geologic events "ridiculous." He did compliment us in a backhand way, saying, "In previous encounters with these authors, I have found them to be good skeptics. Here, however, they are very uncritical of the data they employ and appear to be engaged in statistical nonsense."

The other reviewer, Eugene Shoemaker of the U.S. Geological Survey, delivered quite negative news as well. He also found our hypothesis "naive" and our analysis "misleading." He proceeded to show that our extinction and cratering dates were completely mismatched. As for the galactic model, Shoemaker found that the times of plane crossing were out of phase with those of the mass extinctions and crater data. He was, however, glad to see that I had "gotten religion" on the subject of impacts and mass extinctions. (It is interesting to note that Shoemaker, in his last paper before his tragic death in 1997, came around to supporting the cyclical galactic hypothesis.) We made a long-distance phone call to Philip Campbell at *Nature* in London to assure him that we could revise our manuscript to meet the criticisms of the two reviewers.

So we revised our paper and sent it back to the journal, and again we waited, but this time with even more anxiety. Then, on February 9, I received a phone call from Walter Alvarez. He had apparently gotten wind of our study, probably from a *Nature* reviewer, and wanted to know whether it was similar to work that had been done at Berkeley and that also had been submitted to *Nature*. I was reluctant to talk about an unpublished paper, but I gathered from Alvarez that he and astrophysicist Richard Muller at Berkeley also had analyzed the craters and were postulating periodic comet showers every 28 million years. The Alvarez group was even planning a conference on the new discoveries of periodicity in early March, and he invited us to attend and present our findings.

We received two preprints of the Berkeley work a few days later. Their model of periodicity hypothesized a dwarf companion star to the sun, cycling in a long elliptical orbit that brought it into the Oort cloud every 26 million years. Stothers suspected that such a wide binary star or planet might have an unstable orbit. More critically, this companion star seemed like an extra moving part, not really needed to explain the periodicity of impacts on Earth if one considered that the galaxy itself has a built-in cycle of about 30 million years. We thought it best not to attend the conference, because the papers were all still under review and none had been accepted for publication. In fact, even Raup and Sepkoski's study of periodic extinctions, which had first caught the interest of astrophysicists, had not yet appeared in print.

Word soon came that other groups had submitted manuscripts to *Nature* trying to explain the cosmic periodicity. These would all be discussed at the upcoming meeting at Berkeley. Just to be on the safe side, to protect what we felt was our priority in discovering the evidence for comet showers, we sent copies of our paper to Dick Kerr at *Science* and Cheryl Simon at *Science News*, and we hand delivered one to Walter Sullivan at the *New York Times*, but all with the understanding that no release should be made until the article was scheduled for publication. We wanted them to be aware of our work, should they learn of the other papers that had been submitted to *Nature*.

Events on the West Coast dictated our next actions. Stothers received a call from Muller, who told him that he would be presenting a seminar about the companion star hypothesis at the Berkeley Physics Department. Muller invited David Perlman of the *San Francisco Chronicle* to report on the new discoveries—it seems that an enterprising Berkeley student had tried to sell the story to the *Chronicle*, forcing Muller to go public. Although we reminded Muller of *Nature*'s ban on making a public announcement, he countered that the press would publish the story anyway.

When we spoke with *Nature*'s editorial department in London, they advised us, considering the circumstances, to go ahead with our own release to the press. I called Sullivan and explained the situation, and he read the paper and decided to print something on it. A brief note appeared in the Sunday edition of the *New York Times* on February 19, 1984. Monday's *San Francisco Chronicle* contained a spread by Perlman, mainly on the Berkeley research, and the Tuesday *Science Times* featured a story on both the companion star and galactic oscillation hypotheses.

Even with these leaks to the press, the appearance of the comet shower papers in *Nature* caused quite a stir. All the papers were published in the April 19, 1984, issue—with a photograph of Meteor Crater and the words "Mass Extinctions" emblazoned on the cover—in the order in which they had been received by the journal. This showed some interesting timing. After our lead paper on periodic comet showers, impacts, and the sun's galactic oscillation (received on November 16, 1983) came a paper by physicists Richard Schwartz and Philip James of the University of Missouri, received just one day later, also suggesting that the sun's oscillations led to the extinctions, but not through impacts. Schwartz and James proposed that the extinctions were caused by increased levels of cosmic rays when the solar system was at the extremes of its vertical oscillations. Because this idea did not account for the periodic impacts, and because the timing of the extinctions and the maximum excursions from the galactic plane were quite far off, the hypothesis did not arouse much interest.

After the galactic oscillation papers came two companion star papers (which, incredibly, arrived at the journal offices on the same day, January 3, 1984). The first was written by physicists Dan Whitmire of the University of Southwestern Louisiana and Al Jackson of Computer Services Corporation in Houston, and the second by Muller and Marc Davis of Berkeley, and Piet Hut of Princeton's Institute for Advanced Study. The Berkeley group proposed the name Nemesis for the companion star, and this caught on with scientists and the media. The last paper, received by the journal on January 30, was Alvarez and Muller's spectral analysis of impact crater ages, giving a period of 28 million years.

The same issue of *Nature* in which our paper appeared included an editorial by Editor in Chief John Maddox, and a "News and Views" commentary by noted British geologist Anthony Hallam. In his editorial, Maddox focused on the problem of circulating preprints and releasing information prior to publication. Hallam, by contrast, concentrated on the reliance of Raup and Sepkoski's results on their choice of geologic timescale. He suggested that, if some other timescale were used (and there were at the time several that differed to some extent), then the periodicity would disappear. But Raup and Sepkoski had already shown, and Stothers and I had confirmed, that the 26 million–year periodicity was present no matter which geologic timescale was used. Changing the timescale amounted to small random changes in the mass-extinction dates, but the periodicity was quite robust to such random variations in the dates of the events.

From the very start, we realized that this would be a subject of great interest to the press. Besides the immediate newspaper and news magazine coverage, Dennis Overbye of *Discover* approached us about a feature article.

FIGURE 11.6 Shiva, Hindu god of periodic destruction and renewal of the world, as portrayed at the Ajanta Caves in India, carved into Deccan basalts.

After interviewing us, Overbye laid out the basics of the galactic oscillation hypothesis. He also spelled out the Nemesis hypothesis, and the potential problems with the stability of such a distant companion star. Later that year, the impact/mass-extinction story even made the cover of *Time*.

My student Bruce Haggerty and I thought that the hypothesis needed a name, so we decided to call it the Shiva hypothesis, after the Hindu god of destruction and renewal (figure 11.6). This matched the idea that extinctions destroy the old world and give rise to a subsequent radiation of new species in the postextinction world. The net result of this research is the potential connection between comet impacts and the dynamics of the Milky Way galaxy. The signal of Earth's oscillation through the galactic disk may be recorded in mass extinctions of life on our planet, like a seismograph detecting earthquakes.

12

• • • •

Geological Upheavals and Dark Matter

Continental drift, rifting, and compression, earthquakes, volcanism,
transgression and regression, and polar wander have undoubtedly
a grandiose causal interconnection. . . . Which, however, is cause
and which is effect, only the future will reveal.

ALFRED WEGENER, *THE ORIGIN OF CONTINENTS AND OCEANS*

Do geologists dream of a final theory? Most people would say that geology
already has its paradigmatic theory in plate tectonics. The discovery of plate
tectonics 50 years ago was one of the great scientific achievements of the
twentieth century, but is plate tectonic theory complete? I think not. Plate
tectonics describes Earth's present geology in terms of the geometry and
interactions of the surface plates. Plate motions can be extrapolated back
in time (and into the future), but one cannot yet derive the history of plate
tectonics from first principles. The history can be interpreted post hoc, but
an important question remains: Why did geologic events like hotspot volcanism, fluctuations of seafloor spreading, tectonic episodes, and breakup
of continents occur exactly when and where they did? Are they random,
or do they follow some sort of a pattern in time or space? A complete theory of Earth should explain geologic activity in the spatial domain, as plate
tectonics does quite well for the present (once you incorporate hotspots into
the paradigm), but also in the time and frequency domains. Recent findings suggest to me that geology may be on the threshold of a new theory
that seeks to explain Earth's geologic activity in time and space in the context of its astronomical surroundings.

In late 1983, after we submitted our "galactic carousel" paper to *Nature*,
Richard Stothers and I began work on a follow-up article, intended for *Science*, reporting on a proposed ubiquity of an approximately 30 million–year
cycle in various aspects of the geologic record. I picked up this trail after

reading a paper by Al Fischer and Mike Arthur at Princeton, published in 1977, that suggested a 32 million–year cycle in marine biological diversity and oceanic climate, which they thought might be related to tectonics. The idea of a roughly 30 million–year cycle of geologic events has a long history in the geological literature. In the early twentieth century, Amadeus W. Grabau, the great student of sedimentary strata, proposed periodic fluctuations of sea level, driven by tectonic activity and mountain building with an approximately 30 million–year cycle. In the 1920s, the noted British geologist Arthur Holmes, armed with a few radiometric age determinations, suggested a similar 30 million–year period in Earth's tectonic activity. But the idea of cycles in the geologic record fell out of favor, and most geologists rejected the subject as simply our propensity for seeing cycles where there are none. None of these early papers performed statistical analysis of the geologic events. It was assumed that the dating of the events was too crude for quantitative analysis.

A number of papers over the years have connected fluctuations in sea level and global climate to changes in the plate tectonic regime. For example, variations in the rates of seafloor spreading could potentially control long-term sea level. Rapid spreading along the mid-ocean ridges produces more hot, young ocean crust. This hot crust is buoyant, and so the ridge crest is uplifted, displacing water from the ocean basins to form shallow seas on the continents. When ocean floor spreading slows down, the new crust has time to cool and subside, meaning ocean depths are greater, and the waters of the continental seas can drain back into the ocean basins. At these times, the continents stand high and dry. Marine biodiversity might be expected to decrease when sea level is low and continental shelf areas are exposed above water.

The potential climate connection comes from the fact that water-covered areas absorb much more solar radiation than exposed land. The surface ocean heats up, and water can store a lot of heat energy. The air above the water is warmed, and the climate becomes quite balmy. Conversely, bare, exposed continents reflect much of the solar radiation back to space. The heat energy that is absorbed quickly reradiates into space and is lost. Climate can be much cooler and drier, with seasonal extremes in hot and cold temperatures. Furthermore, global climate also depends critically on the amount of carbon dioxide in the atmosphere. At times of rapid seafloor spreading and subduction, more carbon dioxide is released by volcanism, producing a greater greenhouse effect and hence a warmer global climate.

If the global sea level and climate were fluctuating with a 30 million–year cycle, it made sense to ask whether global tectonics, which apparently controls long-term climate, was varying with the same cycle. In fact, geologist Paul Damon at the University of Arizona published a paper in 1970 proposing that global tectonics and sea level varied with a mean periodicity of about 36 million years. In 1984, I made a search of the geological literature and compiled the best data sets I could find of sea level, mountain building, and various kinds of volcanism. When we analyzed these phenomena with the same techniques used on the mass extinctions and impact craters, Stothers and I found that, indeed, an approximately 30 million–year cycle seems to run through these records of geologic upheavals. We also gathered information on the times of changes in the speed or direction of spreading of the ocean crust. This data set of variations in plate motions also seemed to show a 30 million–year cycle, but the number of events available for analysis was quite small.

Most geologists believe that the geologic record preserves the workings of an essentially random system. Almost no one considers that the record might be periodic on a grand scale. This idea of random processes has permeated geologic thought from before Charles Lyell and down to the present day. When Stothers and I first proposed that Earth's geologic processes may be behaving periodically, in our *Science* article in late 1984, we got a chilly reception. This was due, in part, to the many papers over the years claiming to find one period or another in the geologic record, which did not survive close scrutiny.

Other scientists said that our database was too small, or covered too few cycles, to produce meaningful statistics showing that the geologic cycles were real. But I also believe that accepting the possibility of long-term, regular geologic cycles goes against the Lyellian bias of many geologists against cyclicity in Earth's history. Take, for example, mathematician Milutin Milankovitch's (1879–1958) early-twentieth-century theory to explain the cyclical climate and sea-level fluctuations during the ice age as a result of small changes in Earth's orbit and tilt, caused by the gravitational effects of the moon and planets. This hypothesis languished for years. After all, the idea proposed the interrelatedness of Earth sciences and celestial mechanics, smacked of astrology, and violated two of Lyell's laws. The Milankovitch hypothesis became widely accepted only in the mid-1970s, after a convincing quantitative study of the geologic record of climate and sea level by climate scientists Jim Hays, John Imbrie, and Nick Shackleton, published in *Science*, showed that the climatic cycles matched up beautifully with the proposed forcing through orbital dynamics. Milankovitch was eventually posthumously honored for his groundbreaking work in climatology.

What could be driving the long-term changes in volcanism, tectonics, sea level, and climate at such regular, if widely spaced, intervals? At first, Stothers and I thought that the periodic impacts might somehow be affecting deep-seated geologic processes.

I suggested, in a short note in *Nature* in 1987, that large impacts may so deeply excavate and fracture the crust (to depths in excess of 20 kilometers [12 miles]) that the reduction in pressure in the upper mantle would result in large-scale melting. This would lead to the production of flood basalts, and possibly create a mantle hotspot at the site of the impact. Hotspots lead to continental breakup, which can cause increased tectonics and changes in ocean floor–spreading rates. Unfortunately, no known terrestrial impact structure has a clear association with flood basalt volcanism, although some Martian volcanic outpourings seem to be located along radial and concentric fractures related to large impacts.

It may be that the massive earthquake waves produced by the impact, traveling across and through Earth, could set off volcanism in distant places. Some have suggested that the impact at Chicxulub, 66 million years ago, would have sent seismic surface waves racing around Earth to focus at its exact antipode in the Indian Ocean. There, at the antipodal point, the seismic energy may have been concentrated enough to fracture the crust, leading to Deccan Trap volcanism. Something similar apparently happened on the planet Mercury, where the huge impact that produced the Caloris basin on one side of Mercury lies exactly opposite an area of fractured chaotic terrain. On Mars, some volcanic edifices occur near the antipodes of large impact basins. The problem with this idea with regard to the

Chicxulub impact and Deccan Traps is that, according to reconstructions of the continental positions 66 million years ago, the Yucatán was at about 20 degrees north latitude and India at the time was at about 20 degrees south latitude, sitting over the Réunion hotspot, but the two places were not 180 degrees apart in longitude.

Another possibility, however, involves so-called PKP seismic waves. These PKP body waves, generated by an impact or earthquake and moving through Earth's interior, are refracted by Earth's core and are known to concentrate at 144 degrees from the impact point. This is approximately the longitudinal separation between Chicxulub and the Réunion hotspot/Deccan Traps at the time of their eruption during the Late Cretaceous. So a large impact on one part of the globe may be able to trigger or enhance volcanic eruptions on the other side of the world.

In the early 1990s, I went back to the library at New York University and searched the major journals page by page for data sets related to geologic changes in sea level, tectonics, various kinds of volcanism, discontinuities in seafloor-spreading rates, and indicators of ancient climates, such as the presence of oceans depleted in oxygen or the occurrence of salt deposits indicating a hot, dry climate. I was able to recognize 77 such documented events in Earth's history over the past 260 million years (figure 12.1).

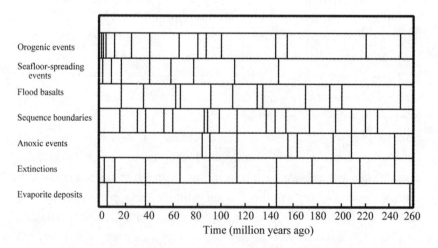

FIGURE 12.1 The spectral distribution of 77 geologic events over the past 260 million years.

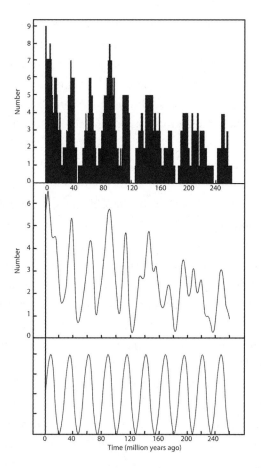

FIGURE 12.2 A plot of the 77 geologic events over the past 260 million years, including mass extinctions, tectonic episodes, changes in seafloor spreading, flood basalt volcanism, and sea-level and climate change. Comparison of running averages (with a 5 million–year moving window; *top*) and smoothed data (*center*) with a 26 million–year cycle (*bottom*) shows that much of the variability can be attributed to that cycle.

My student Ken Caldeira and I analyzed the revised data and found a strong 26 to 27 million–year period of repetition. Again, we saw the rhythm of the geologic upheavals (figure 12.2). Stothers did the same for geomagnetic reversals and detected an approximately 30 million–year periodicity. What could be causing these geologic cycles?

The potential key to this conundrum may come from astrophysics, in the form of invisible dark matter. Remember that during its cosmic orbit, every 30 million years or so, Earth is suggested to pass through clumps of dark matter concentrated along the galactic midplane. In 1986, astrophysicist Lawrence Krauss and Nobel Prize winner Frank Wilczek of Harvard University, and (independently) Katherine Freese at the Harvard-Smithsonian Center for Astrophysics, proposed that these dark matter particles could be captured by Earth, where they would accumulate in the planet's core. While inside the core, the number density of dark matter particles could become great enough that they would undergo mutual annihilation (the particle could be its own antiparticle), producing prodigious amounts of heat in Earth's interior.

I tracked down a paper published in 1998 in the journal *Astroparticle Physics* (which no geologist would ever read), by Indian astrophysicists Asfar Abbas and Samar Abbas (father and son) working at Utkal University, that provided a potential missing link. They, too, were interested in dark matter and its interactions with the planet. They calculated the amount of energy released in the annihilation of dark matter captured by Earth as it passed through a dense clump of dark matter and found that mutual annihilation among the weakly interacting massive particles could produce more than 10^{16} watts. This is 1,000 times greater than Earth's present heat flow (about 10^{13} watts), and much greater than the estimated power required in Earth's interior to generate the planet's magnetic field (about 10^{11} watts). Putting the expected 30 million–year periodicity in encounters with clumps of dark matter together with the effects of Earth's capture of this unstable matter produces a plausible hypothesis with regard to the origin of pulses of geologic activity.

Earth's mantle may be inherently unstable. Excess heat from the core can raise the temperature of the so-called D double-prime boundary layer at the base of the mantle. Such a pulse of heat can create a mantle plume, a rising column of hot mantle rock with a broad head and narrow tail. When these rising plumes impinge upon Earth's crust, they create a hotspot, initiate flood basalt eruptions, and commonly lead to continental fracturing and the initiation of seafloor spreading. The new source of periodic heating in the interior of our planet, dark matter, could lead to periodic outbreaks of mantle plume activity and changes in convection patterns in Earth's core and mantle, which could affect global tectonics, volcanism, geomagnetic field reversals, climate, and life, such as we have seen happen in the past. The new hypothesis links geologic events on Earth with the structure and dynamics of the Milky Way galaxy.

It is still too early to tell whether the ingredients of this hypothesis will withstand further examination and testing. Of course, correlations among geologic events can occur even if they are not part of a periodic pattern, and long-term geologic cycles may exist apart from any cosmic connections. The virtue of the galactic explanation for terrestrial periodicity lies in its universality (because all stars in the galaxy's disk, with their planets, undergo a similar vertical oscillation) and in its linkage of biological and geologic evolution on Earth, and perhaps in other solar systems, with the great cycles of the galaxy. The old battle between gradualist versus catastrophist views of Earth's history may have only been shadow boxing. Both types of change seem to have been important in creating the geologic record we see today. Cataclysms, such as impacts of large asteroids and comets or massive flood basalt volcanism, are as much a part of Earth's history as the slow motions of plate tectonics or the steady erosion by water, wind, and glacial ice.

Epilogue

WHAT DOES IT ALL MEAN? A NEW GEOLOGY

The new understanding of the terrestrial record can thus lead to a
new understanding of the state and evolution of the galaxy itself.

VICTOR CLUBE AND WILLIAM NAPIER, "GIANT COMETS
AND THE GALAXY"

I have tried, in these pages, to present some of what I see as the mounting
evidence that we may be on the threshold of a revolution in geology and
that the future holds the potential for an integration of the geological and
astronomical sciences. In this view, the geological sciences can be regarded
as a sub-branch of astrophysics, with connections to larger questions of the
state of the solar system and the Milky Way galaxy. To understand our
planet, one must study Earth in its cosmic context. The timeworn focus of
geologists on what goes on inside and upon the surface of Earth must be
balanced by an appreciation of larger-scale events beyond our planet.

Einstein said that the only way to make progress in a subject is to question its axioms. The axiomatic concept of uniformitarianism in geology,
as proposed by Charles Lyell 186 years ago, has far outlived its usefulness.
The three laws of Lyell—that all geologic processes are gradual, that extraterrestrial forces are not to be invoked, and that there are no astronomical
cycles which control Earth's destiny—were advanced, as we now see,
largely on theoretical and theological grounds. Lyell's ideas grew from his
assuredness that Earth was designed by an intelligent Creator who acted
always with rational and gradual changes. Cataclysmic revolutions were
not allowed; those were the purview of the French, whose revolutionary
ideas in politics had turned the world upside down. It was up to Lyell, an
Englishman, to bring geology back to considering the importance of

gradual geologic changes acting over immensely long periods of time. Cataclysms were to be outlawed.

I was originally one of the doubters of the impact hypothesis. I was still influenced by the ghost of Lyell. But I changed my mind as the evidence for the end-Cretaceous impact at Chicxulub became overwhelming. Given what we now know about the probability of impacts of comets and asteroids of various sizes, and calculations of the effects of those impacts, it seems inescapable that large-body impact events have affected the history of life on our planet. Discovering evidence of impacts is not easy, but we have signs of the potential involvement of impacts in several extinctions, and the record is expected to get better. However, the correlation of some extinction events with cataclysmic flood basalt volcanism and related times of ocean anoxia suggests that internal Earth pressures also can create global environmental disasters capable of devastating life. Our planet is truly beset by catastrophic changes from within and from above.

Large-body impacts and flood basalt volcanism are part of normal processes that affect Earth's history, and they may be linked. Such events are rare, but the long ages of the geologic record ensure that Earth will, from time to time, experience these catastrophes. The continuing reluctance of geologists to move from a one-off impact hypothesis, limited to the impact at the end of the Cretaceous, to a fuller appreciation of the potential importance of astronomy in Earth's history is certainly partly a continuation of the Lyellian paradigm. An almost physical loathing of cataclysms still leads some to reject the idea of sudden changes (especially extraterrestrial ones) out of hand.

Some geologists seem to long for the old days, when a multitude of hypotheses were available to explain mass extinctions, including transgressions or regressions of the sea, movements of the continents, periods of climate cooling and glaciation, or warming of the planet. But the playing field has changed with the coming of the impact hypothesis and the added realization that huge volcanic eruptions also can have great effects on the environment. It is possible that these geologic factors may all be related through the gravitational effects of dark matter in the galaxy causing periodic comet impacts, and possibly through the ability of Earth to capture unstable dark matter, which can heat the planet's interior and cause geologic and volcanic paroxysms from within. It is my contention that it is through its relationship with astronomy that geology will become truly predictive of the past and future.

At present, the idea of Earth's interaction with dark matter is a mixture of hypothesis and conjecture, but if one accepts the premises, the geologic record begins to look very different from what it was purported to be in a Lyellian world. We can no longer depend on the slow and gradual progression of geologic change or describe evolution as the history of life's continuing triumphs over the slow rhythms of geologic upheavals and climatic changes. In the new geology, life, though resilient, is bombarded from space, exposed to massive volcanic eruptions emanating from Earth's deep interior, and suffers periodic catastrophic environmental changes and mass extinctions. These events change the course of evolution. All this activity may be merely side effects of encounters with dark matter during the rhythmic motion of the stars around and through the great cosmic wheel of the galaxy. If this hypothesis bears fruit, then it is an opportunity for a great collaboration of geologists and astronomers. The new geology promises to be cataclysmal, cyclical, and governed in part by astronomical forces. This is clearly a long way from the geology I was taught in school.

Sources and Further Reading

This bibliography is not meant to be comprehensive. It contains sources that I used in writing this book, papers of historical interest, and other references that should allow the reader to delve further into the vast literature on the subjects of this book.

Introduction

Alvarez, L. W., W. Alvarez, F. Asaro, and H. V. Michel. Extraterrestrial cause of the Cretaceous/Tertiary extinction: Experimental results and theoretical interpretation. *Science* 208 (1980): 1095–1108.

Descartes, R. *Meditations on First Philosophy*. Paris: Adam and Tannery, 1641.

Lyell, C. *Principles of Geology, Being an Attempt to Explain the Former Changes of the Earth's Surface by Processes Still in Operation*. 3 vols. London: Murray, 1830–1833.

1. Catastrophism Versus Gradualism

Cuvier, G. *Essay on the Theory of the Earth with Geological Illustrations by Professor Jameson*. 5th ed. Edinburgh: Blackwell, 1827.

Dana, J. D. *Creation; or, the Biblical Cosmogony in the Light of Modern Science*. Oberlin, Ohio: Goodrich, 1885.

Dana, J. D. *The Geological Story Briefly Told*. New York: American Books, 1875.

Dana, J. D. *Manual of Geology*. Taylor, N.Y.: Blakeman, 1863.

Darwin, C. *On the Origin of Species by Means of Natural Selection or the Preservation of Favored Races in the Struggle for Survival*. London: Murray, 1859.

du Toit, A. L. *Our Wandering Continents, an Hypothesis of Continental Drifting.* Edinburgh: Oliver and Boyd, 1937.

Hutton, J. *Theory of the Earth, or an Investigation of Laws Observable in the Composition, Dissolution and Restoration of Land Upon the Globe.* 2 vols. London: Cadell and Davies, 1795.

Paley, W. *Natural Theology, or Evidences of the Existence and Attributes of the Deity Collected from the Appearances of Nature.* London: Faulder, 1802.

Rudwick, M. J. S. *Georges Cuvier, Fossil Bones, and Geological Catastrophes.* Chicago: University of Chicago Press, 1997.

2. Lyell's Laws

Lyell, C. *Life, Letters and Journals of Sir Charles Lyell, Bart.* 2 vols. London: Murray, 1881.

Lyell, C. *Principles of Geology, Being an Attempt to Explain the Former Changes of the Earth's Surface, by Reference to Causes Now in Operation.* 3 vols. London: Murray, 1830–1833.

Sloane, D. S. Evolution—Its meaning. In *Creation by Evolution*, edited by F. Mason. New York: Macmillan, 1928.

Whiston, W. *A New Theory of the Earth, From its Origin to the Consummation of All Things.* London: Roberts, 1696.

3. The Alvarez Hypothesis

Albertao, G. A., and P. P. Martins Jr. A possible tsunami deposit at the Cretaceous-Tertiary boundary in Pernambuco, northeastern Brazil. *Sedimentary Geology* 104 (1996): 189–201.

Alegret, L., E. Thomas, and K. C. Lohmann. End-Cretaceous marine mass extinction not caused by productivity collapse. *Proceedings of the National Academy of Sciences of the United States of America* 109 (2012): 728–32.

Alvarez, L. W. Mass extinctions caused by large bolide impacts. *Physics Today* 40 (1987): 24–33.

Alvarez, L. W., W. Alvarez, F. Asaro, and H. V. Michel. Extraterrestrial cause of Cretaceous/Tertiary extinction: Experimental results and theoretical interpretation. *Science* 208 (1980): 1095–1108.

Alvarez, W. *T. Rex and the Crater of Doom.* Princeton, N.J.: Princeton University Press, 1997.

Alvarez, W., L. Alvarez, F. Asaro, and H. V. Michel. The end of the Cretaceous: Sharp boundary or gradual transition. *Science* 223 (1984): 1183–86.

Alvarez, W., J. Smit, W. Lowrie, F. Asaro, S. V. Margolis, P. Claeys, M. Kastner, and A. R. Hildebrand. Proximal impact deposits at the Cretaceous-Tertiary boundary in the Gulf of Mexico: A restudy of DSDP Leg 77 Sites 536 and 540. *Geology* 20 (1992): 697–700.

Bernaola, G., and S. Monechi. Calcareous nannofossil extinction and survivorship across the Cretaceous-Paleogene boundary at Walvis Ridge (ODP Hole 1262C, South Atlantic Ocean). *Palaeogeography, Palaeoclimatology, Palaeoecology* 255 (2007): 132–56.

Bleiweiss, R. Fossil gap analysis supports early Tertiary origin of trophically diverse avian orders. *Geology* 26 (1998): 323–26.

Bohor, B. F., E. E. Foord, P. J. Modreski, and D. M. Triplehorn. Mineralogic evidence for an impact event at the Cretaceous-Tertiary boundary. *Science* 224 (1984): 867–69.

Bohor, B. F., P. J. Modreski, and E. E. Foord. Shocked quartz in the Cretaceous-Tertiary boundary clays: Evidence for a global distribution. *Science* 236 (1987): 705–9.

Bohor, B. F., D. M. Triplehorn, D. J. Nichols, and H. T. Millard Jr. Dinosaurs, spherules, and the magic layer: A new K-T boundary site in Wyoming. *Geology* 15 (1987): 896–99.

Booth, B., and F. Fitch. *Earth Shock*. New York: Walker, 1979.

Bostwick, J. A., and F. T. Kyte. The size and abundance of shocked quartz in Cretaceous-Tertiary boundary sediments from the Pacific basin. *Geological Society of America Special Papers* 307 (1996): 403–15.

Bourgeois, J., T. A. Hansen, P. L. Wiberg, and E. G. Kauffman. A tsunami deposit at the Cretaceous-Tertiary boundary in Texas. *Science* 241 (1988): 567–70.

Bralower, T. J., C. K. Paull, and R. M. Leckie. The Cretaceous-Tertiary cocktail: Chicxulub impact triggers margin collapse and extensive sediment gravity flows. *Geology* 26 (1998): 331–34.

Carlisle, D. B., and D. R. Braman. Nanometre-size diamonds in the Cretaceous/Tertiary boundary of Alberta. *Nature* 352 (1991): 708–9.

Chao, E. C. T., R. Huttner, and H. Schmidt-Kaler. *Principal Exposures of the Ries Meteorite Crater in Southern Germany*. Munich: Bayerisches Geologisches Landesamt, 1978.

Claeys, P., W. Kiessling, and W. Alvarez. Distribution of Chicxulub ejecta at the Cretaceous-Tertiary boundary. *Geological Society of America Special Papers* 356 (2002): 55–68.

De Laubenfels, M. W. Dinosaur extinction: One more hypothesis. *Journal of Paleontology* 30 (1956): 207–17.

Evans, N. J., and C. F. Chai. The distribution and geochemistry of platinum-group elements as event markers in the Phanerozoic. *Palaeogeography, Palaeoclimatology, Palaeoecology* 132 (1997): 373–90.

Fornaciari, E., L. Giusberti, V. Luciani, F. Tateo, C. Agnini, J. Backman, M. Oddone, and D. Rio. An expanded Cretaceous-Tertiary transition in a pelagic setting of the Southern Alps (central-western Tethys). *Palaeogeography, Palaeoclimatology, Palaeoecology* 255 (2007): 98–131.

Frankel, C. *The End of the Dinosaurs: Chicxulub Crater and Mass Extinctions*. Cambridge: Cambridge University Press, 1999.

Hallam, A. End-Cretaceous mass extinction event: Argument for terrestrial causation. *Science* 238 (1987): 1237–42.

Hildebrand, A. R., and W. V. Boynton. Proximal Cretaceous-Tertiary boundary impact deposits in the Caribbean. *Science* 248 (1990): 843–47.

Hildebrand, A. R., G. T. Penfield, D. A. Kring, M. Pilkington, A. Z. Camargo, S. B. Jacobsen, and W. V. Boynton. Chicxulub crater: A possible Cretaceous/Tertiary boundary impact crater on the Yucatán Peninsula, Mexico. *Geology* 19 (1991): 867–71.

Hildebrand, A. R., M. Pilkington, M. Connors, C. Ortiz-Aleman, and R. E. Chavez. Size and structure of the Chicxulub crater revealed by horizontal gravity gradients and cenotes. *Nature* 376 (2002): 415–17.

Hsü, K. J. *The Great Dying: Cosmic Catastrophe, Dinosaurs, and the Theory of Evolution.* New York: Harcourt Brace Jovanovich, 1986.

Jones, D. S., P. A. Mueller, J. R. Bryan, J. P. Dobson, J. E. T. Channell, J. C. Zachos, and M. A. Arthur. Biotic, geochemical and paleomagnetic changes across the Cretaceous/Tertiary boundary at Braggs, Alabama. *Geology* 15 (1987): 311–15.

Keller, G. Deccan volcanism, the Chicxulub impact and the end-Cretaceous mass extinction: Coincidence? Cause and effect? *Geological Society of America Special Papers* 505 (2014): 57–89.

Keller, G., T. Adatte, Z. Berner, M. Harting, G. Baum, M. Prauss, A. A. Tantawy, and D. Stüben. Chicxulub impact predates K-T boundary: New evidence from Brazos, Texas. *Earth and Planetary Science Letters* 255 (2007): 339–56.

Keller, G., T. Adatte, W. Stinnesback, M. Affolter, L. Schilli, and J. Guadalupe Lopez-Oliva. Multiple spherule layers in the late Maastrichtian of northeastern Mexico. *Geological Society of America Special Papers* 356 (2002): 145–61.

Keller, G., T. Adatte, W. Stinnesbeck, M. Rebolledo-Vieyra, J. Urrutia Fucugauchi, U. Kramar, and D. Stüben. Chicxulub impact predates the K-T boundary mass extinction. *Proceedings of the National Academy of Sciences of the United States of America* 101 (2004): 3753–58.

Keller, G., J. G. Lopez-Oliva, W. Stinnesbeck, and T. Adatte. Age, stratigraphy, and deposition of near-K/T siliciclastic deposits in Mexico: Relation to bolide impact? *Geological Society of America Bulletin* 109 (1997): 410–28.

Keller, G., W. Stinnesbeck, T. Adatte, and D. Stüben. Multiple impacts across the Cretaceous-Tertiary boundary. *Earth-Science Reviews* 62 (2003): 327–63.

Kent, D. V. An estimate of the duration of the faunal change at the Cretaceous-Tertiary boundary. *Geology* 5 (1978): 769–71.

Kiyokawa, S. Cretaceous-Tertiary boundary sequence in the Cacarajicara Formation, western Cuba: An impact-related, high-energy, gravity flow deposit. *Geological Society of America Special Papers* 356 (2002): 125–44.

Koeberl, C., and K. G. MacLeod, eds. Catastrophic events and mass extinctions: Impacts and beyond. *Geological Society of America Special Papers* 356 (2002).

Kousoukos, E. A. An extraterrestrial impact in the early Danian: A secondary K/T boundary event. *Terra Nova* 10 (1998): 68–73.

Krogh, T. E., S. L. Kamo, and B. F. Bohor. Fingerprinting the K/T impact site and determining the time of impact by U-Pb dating of single shocked zircons from distal ejecta. *Earth and Planetary Science Letters* 119 (1993): 425–29.

Krogh, T. E., S. L. Kamo, V. L. Sharpton, L. E. Marin, and A. R. Hildebrand. U-Pb ages of single shocked zircons linking distal K/T ejecta to the Chicxulub crater. *Nature* 366 (1993): 731–34.

Kyte, F. T. A meteorite from the Cretaceous/Tertiary boundary. *Nature* 396 (1998): 237–39.

Longrich, N. R., B.-A. S. Bhullar, and J. A. Gauthier. Mass extinction of lizards and snakes at the Cretaceous-Paleogene boundary. *Proceedings of the National Academy of Sciences of the United States of America* 109 (2012): 21396–401.

Longrich, N. R., T. Tokaryk, and D. J. Field. Mass extinction of birds at the Cretaceous/ Paleogene (K-Pg) boundary. *Proceedings of the National Academy of Sciences of the United States of America* 108 (2011): 15253–57.

Lopez-Oliva, J. G., and G. Keller. Age and stratigraphy of near-K/T boundary silici-clastic deposits in northeastern Mexico. *Geological Society of America Special Papers* 307 (1996): 227–42.

MacLeod, N. *The Great Extinctions: What Causes Them and How They Shape Life.* Buf-falo, N.Y.: Firefly Books, 2013.

Maurrasse, F. J.-M. R. New data on the stratigraphy of the southern peninsula of Haiti. *Transactions du 1er Colloque sur la Geologie D'Haiti, Port-au-Prince, 27–29 March 1980* (1982): 184–98.

Maurrasse, F. J.-M. R., and G. Sen. Impacts, tsunamis, and the Haitian Cretaceous-Tertiary boundary layer. *Science* 252 (1991): 1690–93.

McHone, J. F., R. P. Nieman, C. F. Lewis, and A. M. Yates. Stishovite at the Cretaceous/ Tertiary boundary, Raton, New Mexico. *Science* 243 (1989): 1182–84.

Meyers, P. A., and B. R. T. Simoneit. Global comparisons of organic matter in sedi-ments across the Cretaceous/Tertiary boundary. *Organic Geochemistry* 16 (1989): 641–48.

Morgan, J., M. Warner, J. Brittan, R. Buffler, A. Camargo, G. Christeson, P. Denton, A. Hildebrand, R. Hobbs, H. Macintyre, et al. Size and morphology of the Chicx-ulub impact crater. *Nature* 390 (1997): 472–76.

Morgan, J., M. Warner, and R. Grieve. Geophysical constraints on the size and struc-ture of the Chicxulub impact crater. *Geological Society of America Special Papers* 356 (2002): 39–46.

Nichols, D. J. Selected plant microfossil records of the terminal Cretaceous event in terrestrial rocks, western North America. *Palaeogeography, Palaeoclimatology, Pal-aeoecology* 255 (2007): 22–34.

Nohr-Hansen, H., and G. Dam. Palynology and sedimentology across a new marine Cretaceous-Tertiary boundary section on Nuussuaq, West Greenland. *Geology* 25 (1997): 851–54.

Norris, R. D., and J. V. Firth. Mass wasting of Atlantic continental margins follow-ing the Chicxulub impact event. *Geological Society of America Special Papers* 356 (2002): 79–95.

Ocampo, A. C., K. O. Pope, and A. G. Fischer. Ejecta blanket of the Chicxulub crater from Albion Island, Belize. *Geological Society of America Special Papers* 307 (1996): 75–88.

Ohno, S., T. Kadono, K. Kurosawa, T. Hamura, T. Sakaiya, K. Shigemori, Y. Hiron-aka, T. Sano, T. Watari, K. Otani, et al. Production of sulphate-rich vapour dur-ing Chicxulub impact and implications for ocean acidification. *Nature Geoscience* 7 (2014): 279–82.

Olsson, R. K., K. G. Miller, J. V. Browning, J. D. Wright, and B. S. Cramer. Sequence stratigraphy and sea-level change across the Cretaceous-Tertiary boundary on the New Jersey passive margin. *Geological Society of America Special Papers* 356 (2002): 97–108.

Opik, E. J. On the catastrophic effects of collisions with celestial bodies. *Irish Astronomical Journal* 5 (1958): 34–36.

Paul, C. R. C. Interpreting bioevents: What exactly did happen to planktonic foraminifers across the Cretaceous-Tertiary boundary? *Palaeogeography, Palaeoclimatology, Palaeoecology* 224 (2005): 291–310.

Pierazzo, E., A. N. Hahmann, and L. C. Sloan. Chicxulub and climate: Radiative perturbations of impact-produced S-bearing gases. *Astrobiology* 3 (2003): 99–118.

Pope, K. O. Impact dust not the cause of the Cretaceous-Tertiary mass extinction. *Geology* 30 (2002): 99–102.

Pope, K. O., A. C. Ocampo, and C. E. Duller. Mexican site for K/T impact crater? *Nature* 351 (1991): 105.

Pospichal, J. J. Calcareous nannofossils at the K-T boundary, El Kef: No evidence for stepwise, gradual, or sequential extinctions. *Geology* 22 (1994): 99–102.

"Possible Yucatan Impact Basin." *Sky and Telescope*, March 1982, 249–250.

Powell, J. L. *Four Revolutions in the Earth Sciences: From Heresy to Truth.* New York: Columbia University Press, 2015.

Powell, J. L. *Night Comes to the Cretaceous: Dinosaur Extinction and the Transformation of Modern Geology.* New York: Freeman, 1998.

Rampino, M. R., and R. C. Reynolds. Clay mineralogy of the Cretaceous-Tertiary boundary clay. *Science* 219 (1983): 495–98.

Renne, P. R., A. L. Deino, F. J. Hilgen, K. F. Kuiper, D. F. Mark, W. S. Mitchell III, L. E. Morgan, R. Mundil, and J. Smit. Time scales of critical events around the Cretaceous-Paleogene boundary. *Science* 339 (2013): 684–87.

Ryder, G., D. Fastovsky, and S. Gartner, eds. The Cretaceous-Tertiary event and other catastrophes in Earth history. *Geological Society of America Special Papers* 307 (1996).

Schulte, P., L. Alegret, I. Arenillas, J. A. Arz, P. J. Barton, P. R. Brown, T. J. Bralower, G. L. Christeson, P. Claeys, C. C. Cockell, et al. The Chicxulub asteroid impact and mass extinction at the Cretaceous-Paleogene boundary. *Science* 327 (2010): 1214–18.

Sharpton, V. L., G. B. Dalrymple, L. E. Marin, G. Ryder, B. C. Schuraytz, and J. Urrutia-Fucugauchi. New links between the Chicxulub impact structure and the Cretaceous/Tertiary boundary. *Nature* 359 (1992): 819–21.

Sharpton, V. L., and P. D. Ward, eds. Global catastrophes in Earth history. *Geological Society of America Special Papers* 247 (1990).

Sheehan, P. M., D. E. Fastovsky, R. G. Hoffmann, C. B. Berghaus, and D. L. Gabriel. Sudden extinction of the dinosaurs: Latest Cretaceous, Upper Great Plains, USA. *Science* 254 (1991): 835–39.

Sigurdsson, H., S. D'Hondt, M. A. Arthur, T. J. Bralower, J. C. Zachos, and M. Channell. Glass from the Cretaceous/Tertiary boundary in Haiti. *Nature* 349 (1991): 482–87.

Sigurdsson, H., S. D'Hondt, and S. Carey. The impact of the Cretaceous/Tertiary bolide on evaporite terrane and generation of sulfuric acid aerosols. *Earth and Planetary Science Letters* 109 (1992): 543–59.

Silver, L. T., and P. H. Schultz, eds. Geological implications of impacts of large asteroids and comets on the Earth. *Geological Society of America Special Papers* 190 (1982).

Smit, J., and J. Hertogen. An extraterrestrial event at the Cretaceous-Tertiary boundary. *Nature* 285 (1980): 198–200.

Smit, J., and G. Klaver. Sanidine spherules at the Cretaceous-Tertiary boundary indicate large impact event. *Nature* 292 (1981): 47–49.

Smit, J., and F. T. Kyte. Siderophile-rich magnetic spheroids from the Cretaceous-Tertiary boundary in Umbria, Italy. *Nature* 310 (1984): 403–5.

Smit, J., A. Montanari, N. H. M. Swinburne, W. Alvarez, A. R. Hildebrand, S. V. Margolis, P. Claeys, W. Lowrie, and F. Asaro. Tektite-bearing, deep-water clastic unit at the Cretaceous-Tertiary boundary in northeastern Mexico. *Geology* 20 (1992): 99–103.

Smit, J., Th. B. Roep, W. Alvarez, A. Montanari, P. Claeys, J. M. Grajales-Nishimura, and J. Bermudez. Coarse-grained clastic sandstone complex at the K/T boundary around the Gulf of Mexico: Deposition by tsunami waves induced by the Chicxulub impact? *Geological Society of America Special Papers* 307 (1996): 151–82.

Stanley, S. M. Delayed recovery and the spacing of major extinctions. *Paleobiology* 16 (1990): 401–14.

Stinnesbeck, W., and G. Keller. K/T boundary coarse-grained siliciclastic deposits in northeastern Mexico and northeastern Brazil: Evidence for mega-tsunami or sea-level changes? *Geological Society of America Special Papers* 307 (1996): 197–209.

Swisher, C. C., III, J. M. Grajales-Nishimura, A. Montanari, S. V. Margolis, P. Claeys, W. Alvarez, P. Renne, E. Cedillo-Pardoa, F. J. Maurrasse, G. H. Curtis, et al. Coeval $^{40}Ar/^{39}Ar$ ages of 65.0 million years ago from Chicxulub crater melt rock and Cretaceous-Tertiary boundary tektites. *Science* 257 (1992): 954–58.

Urey, H. C. Cometary collisions and geological periods. *Nature* 242 (1973): 32–33.

Vajda, V. V., and S. McLoughlin. Extinction and recovery patterns of the vegetation across the Cretaceous-Palaeogene boundary—A tool for unraveling the causes of the end-Permian mass-extinction. *Review of Palaeobotany and Palynology* 144 (2007): 99–112.

Vajda, V. V., and S. McLoughlin. Fungal proliferation at the Cretaceous-Tertiary boundary. *Science* 303 (2004): 1489.

Vajda, V., J. I. Raine, and C. J. Hollis. Indication of global deforestation at the Cretaceous-Tertiary boundary by New Zealand fern spike. *Science* 294 (2001): 1700–1702.

Vellekoop, J., S. Esmeray-Senlet, K. G. Miller, J. V. Browning, A. Sluijs, B. van de Schootbrugge, J. S. Sinninghe Damsté, and H. Brinkhuis. Evidence for Cretaceous-Paleogene boundary bolide impact winter conditions from New Jersey, USA. *Geology* 44 (2016): 619–22.

Widmark, J. G. V., and B. Malmgren. Benthic foraminiferal change across the Cretaceous-Tertiary boundary in the deep sea; DSDP Sites 525, 527, and 465. *Journal of Foraminiferal Research* 22 (1992): 81–113.

Witts, J. D., R. J. Whittle, P. B. Wignall, J. A. Crame, J. E. Francis, R. J. Newton, and V. C. Bowman. Macrofossil evidence for a rapid and severe Cretaceous-Paleogene mass extinction in Antarctica. *Nature Communications* 7 (2016). doi:10.1038/11738.

4. Mass Extinctions

Bailer-Jones, C. A. L. The evidence for and against astronomical impacts on climate change and mass extinctions: A review. *International Journal of Astrobiology* 8 (2009): 213–39.

Bambach, R. K. Phanerozoic biodiversity mass extinctions. *Annual Review of Earth and Planetary Sciences* 34 (2006): 127–55.

Benton, M. J. Diversification and extinction in the history of life. *Science* 268 (1995): 52–58.

Benton, M. J. *The Fossil Record 2.* London: Chapman & Hall, 1993.

Darwin, C. *On the Origin of Species by Means of Natural Selection or the Preservation of Favored Races in the Struggle for Survival.* London: Murray, 1859.

Erwin, D. H. *Extinction: How Life on Earth Nearly Ended 250 Million Years Ago.* Princeton, N.J.: Princeton University Press, 2006.

Fischer, A. G., and M. A. Arthur. Secular variations in the pelagic realm. *Society of Economic Paleontologists and Mineralogists Special Publication* 25 (1977): 19–50.

Fox, W. T. Harmonic analysis of periodic extinctions. *Paleobiology* 13 (1987): 257–71.

Hsü, K. J. Sedimentary petrology and biologic evolution. *Journal of Sedimentary Petrology* 56 (1983): 729–32.

Lyell, C. *Principles of Geology, Being an Attempt to Explain the Former Changes of the Earth's Surface by Processes Still in Operation.* 3 vols. London: Murray, 1830–1833.

Melott, A. L., and R. K. Bambach. Do periodicities in extinction—with possible astronomical connections—survive a revision of the geological timescale? *Astrophysical Journal* 773 (2013): 6–11.

Morgan, T. H. *A Critique of the Theory of Evolution.* Princeton, N.J.: Princeton University Press, 1916.

Newell, N. D. Periodicity in invertebrate evolution. *Journal of Paleontology* 26 (1952): 371–85.

Rampino, M. R. Are marine and nonmarine extinctions correlated? *Eos, Transactions of the American Geophysical Union* 69 (1988): 889–95.

Rampino, M. R., and K. Caldeira. Periodic impact cratering and extinction events over the last 260 million years. *Monthly Notices of the Royal Astronomical Society* 454 (2015): 3480–84.

Rampino, M. R., B. M. Haggerty, and T. C. Pagano. A unified theory of impact crises and mass extinctions: Quantitative tests. *Annals of the New York Academy of Sciences* 822 (1997): 403–31.

Raup, D. M. Biogeographic extinction: A feasibility test. *Geological Society of America Special Papers* 190 (1982): 277–81.

Raup, D. M. Impact as a general cause of extinction: A feasibility test. *Geological Society of America Special Papers* 247 (1990): 27–32.

Raup, D. M. A kill curve for Phanerozoic marine species. *Paleobiology* 17 (1991): 37–48.

Raup, D. M. The role of extinction in evolution. *Proceedings of the National Academy of Sciences of the United States of America* 91 (1994): 6758–63.

Raup, D. M. Size of the Permo-Triassic bottleneck and its evolutionary implications. *Science* 206 (1979): 217–18.

Raup, D. M., and J. J. Sepkoski Jr. Periodic extinctions of families and genera. *Science* 231 (1986): 833–36.

Raup, D. M., and J. J. Sepkoski Jr. Periodicity of extinctions in the geologic past. *Proceedings of the National Academy of Sciences of the United States of America* 81 (1984): 801–5.

Raup, D. M., and J. J. Sepkoski Jr. Testing for periodicity of extinctions. *Science* 241 (1988): 94–99.

Sepkoski, J. J., Jr. *A Compendium of Fossil Marine Animal Families.* Milwaukee Public Museum Contributions in Biology and Geology 51. Milwaukee: Milwaukee Public Museum, 1982.

Sepkoski, J. J., Jr. *A Compendium of Fossil Marine Animal Families.* 2nd ed. Milwaukee Public Museum Contributions in Biology and Geology 83. Milwaukee: Milwaukee Public Museum, 1992.

Sepkoski, J. J., Jr. *A Compendium of Fossil Marine Animal Genera.* Bulletin of American Paleontology 363. Ithaca, N.Y.: Paleontological Research Institution, 2002.

Sepkoski, J. J., Jr. Extinction and the fossil record. *Geotimes* 39 (1994): 15–17.

Sepkoski, J.J., Jr. A kinetic model of Phanerozoic taxonomic diversity II: Early Phanerozoic families and multiple equilibria. *Paleobiology* 5 (1979): 222–52.

Sepkoski, J. J., Jr. Patterns of Phanerozoic extinctions: A perspective from global databases. In *Global Events and Event Stratigraphy in the Phanerozoic,* edited by O. H. Walliser, 35–52. Berlin: Springer, 1996.

Sepkoski, J. J., Jr. Ten years in the library: New data confirm paleontological patterns. *Paleobiology* 19 (1993): 43–51.

Sepkoski, J. J., Jr., and D. M. Raup. Was there a 26-Myr periodicity of extinctions? *Nature* 321 (1986): 535–36.

Stigler, S. M., and M. J. Wagner. A substantial bias in nonparametric tests for periodicity in geophysical data. *Science* 238 (1984): 940–45.

Stothers, R. B. Structure and dating errors in the geologic time scale and periodicity in mass extinctions. *Geophysical Research Letters* 16 (1989): 119–22.

5. Kill Curves and Strangelove Oceans

Adushkin, V. V., and I. V. Nemchinov. Consequences of impacts of cosmic bodies on the surface of the Earth. In *Hazards Due to Comets and Asteroids,* edited by T. Gehrels, 721–78. Tucson: University of Arizona Press, 1994.

Alvarez, L. W. Experimental evidence that an asteroid impact led to the extinction of many species 65 million years ago. *Proceedings of the National Academy of Sciences of the United States of America* 80 (1983): 627–42.

Beerling, D. J., B. H. Lomax, D. L. Royer, G. R. Upchurch Jr., and L. R. Kump. An atmospheric pCO_2 reconstruction across the Cretaceous-Tertiary boundary from leaf megafossils. *Proceedings of the National Academy of Sciences of the United States of America* 99 (2002): 7836–40.

Caldeira, K., and M. R. Rampino. The aftermath of the K/T boundary mass extinction: Biogeochemical stabilization of the carbon cycle and climate. *Paleoceanography* 8 (1993): 515–25.

Caldeira, K., M. R. Rampino, T. Volk, and J. C. Zachos. Biogeochemical modeling at mass extinction boundaries: Atmospheric carbon dioxide and ocean alkalinity at the K/T boundary. In *Global Bioevents: Abrupt Changes in the Global Biota Through Time*, edited by E. G. Kaufman and O. H. Walliser, 333–45. Berlin: Springer, 1990.

Chapman, C. R., and D. Morrison. Impacts on the Earth by asteroids and comets: Assessing the hazard. *Nature* 367 (1994): 33–40.

Covey, C., S. J. Ghan, J. J. Walton, and P. R. Weissman. Global environmental effects of impact-generated aerosols: Results from a general circulation model. *Geological Society of America Special Papers* 247 (1990): 263–70.

Covey, C., S. L. Thompson, P. R. Weissman, and M. C. MacCracken. Global climatic effects of atmospheric dust from an asteroid or comet impact on Earth. *Global and Planetary Change* 9 (1994): 263–73.

Croft, S. K. A first-order estimate of shock heating and vaporization in oceanic impacts. *Geological Society of America Special Papers* 190 (1982): 143–51.

Davies-Vollum, K. S., L. D. Boucher, P. Hudson, and A. Y. Proskurowski. A Late Cretaceous coniferous woodland from the San Juan Basin, New Mexico. *Palaios* 26 (2011): 89–98.

Durda, D. D., and D. A. Kring. Ignition threshold for impact-generated fires. *Journal of Geophysical Research* 109 (2004): E08004.

Eldredge, N., and S. J. Gould. Punctuated equilibria: An alternative to phyletic gradualism. In *Models in Paleobiology*, edited by T. J. M. Schopf, 82–115. San Francisco: Freeman, 1972.

Galeotti, S., H. Brinkhuis, and M. Huber. Records of post–Cretaceous-Tertiary boundary millennial-scale cooling from western Tethys: A smoking gun for the impact-winter hypothesis. *Geology* 32 (2004): 529–32.

Gerstl, S. A., and A. Zardecki. Reduction of photosynthetically active radiation under extreme stratospheric aerosol loads. *Geological Society of America Special Papers* 190 (1982): 201–10.

Goldin, T. J., and H. J. Melosh. Self-shielding of thermal radiation by Chicxulub impact ejecta: Firestorm or fizzle. *Geology* 37 (2009): 1135–38.

Griffis, K., and D. J. Chapman. Survival of phytoplankton under prolonged darkness: Implications for the Cretaceous-Tertiary boundary darkness hypothesis. *Palaeogeography, Palaeoclimatology, Palaeoecology* 67 (1988): 305–14.

Hollander, D. J., J. A. McKenzie, and K. J. Hsü. Carbon isotope evidence for unusual plankton blooms and fluctuations of surface water CO_2 in Strangelove Ocean after terminal Cretaceous event. *Palaeogeography, Palaeoclimatology, Palaeoecology* 104 (1993): 229–37.

Hsü, K. J., Q. He, J. A. McKenzie, H. Weissert, K. Perch-Nielsen, H. Oberhänsli, K. Kelts, J. Labrecque, L. Tauxe, U. Krähenbühl, et al. Mass mortality and its environmental and evolutionary consequences. *Science* 216 (1992): 249–56.

Hsü, K. J., and J. A. McKenzie. Carbon-isotope anomalies at era boundaries: Global catastrophes and their ultimate cause. *Geological Society of America Special Papers* 247 (1990): 61–69.

Hsü, K. J., H. Oberhänsli, J. Y. Gao, S. Shu, C. Haihong, and U. Krähenbühl. Strangelove Ocean before the Cambrian explosion. *Nature* 316 (1985): 809–11.

Izett, G. A., W. A. Cobban, J. D. Obradovich, and M. J. Kunk. The Manson impact structure: $^{40}Ar/^{39}Ar$ age and its distal impact ejecta in the Pierre Shale in southeastern South Dakota. *Science* 262 (1993): 729–32.

Jansa, L. F. Cometary impacts into ocean: Their recognition and the threshold constraint for biological extinctions. *Palaeogeography, Palaeoclimatology, Palaeoecology* 104 (1993): 271–86.

Katongo, C., C. Koeberl, B. J. Witzke, R. H. Hammond, and R. R. Anderson. Geochemistry and shock petrography of the Crow Creek Member, South Dakota, USA: Ejecta from the 74-Ma Manson impact structure. *Meteoritics & Planetary Science* 39 (2004): 31–51.

Kring, D. A. The Chicxulub impact event and its environmental consequences at the Cretaceous-Tertiary boundary. *Palaeogeography, Palaeoclimatology, Palaeoecology* 255 (2007): 4–21.

Kring, D. A., and D. D. Durda. Trajectories and distribution of material ejected from Chicxulub impact crater: Implications for post-impact wildfires. *Journal of Geophysical Research* 107 (2002): 6-1–22.

Kyte, F. T. The extraterrestrial component in marine sediments: Description and interpretation. *Paleoceanography* 3 (1988): 235–47.

Kyte, F. T., Z. Zhou, and J. T. Wasson. High noble metal concentrations in a late Pliocene sediment. *Nature* 292 (1981): 417–20.

Kyte, F. T., Z. Zhou, and J. T. Wasson. New evidence on the size and possible effects of a late Pliocene asteroid impact. *Science* 241 (1988): 63–65.

Melosh, H. J. *Impact Cratering: A Geologic Process.* New York: Oxford University Press, 1989.

Melosh, H. J. The mechanics of large meteoroid impacts in the Earth's oceans. *Geological Society of America Special Papers* 190 (1982): 121–27.

Melosh, H. J., N. M. Schneider, K. J. Zahnle, and D. Latham. Ignition of global wildfires at the Cretaceous/Tertiary boundary. *Nature* 343 (1990): 251–54.

Melott, A. L., B. C. Thomas, G. Dreschhoff, and C. K. Johnson. Cometary airbursts and atmospheric chemistry: Tunguska and a candidate Younger Dryas event. *Geology* 38 (2010): 355–58.

Morrison, D., C. R. Chapman, and P. Slovic. The impact hazard. In *Hazards Due to Asteroids and Comets*, edited by T. Gehrels, 59–91. Tucson: University of Arizona Press, 1994.

O'Keefe, J., and T. J. Ahrens. Impact production of CO_2 by the Cretaceous/Tertiary extinction bolide and resultant heating of the Earth. *Nature* 338 (1989): 247–49.

Pierazzo, E., D. A. Kring, and H. J. Melosh. Hydrocode simulation of the Chicxulub impact event and the production of climatically active gases. *Journal of Geophysical Research* 103 (1998): 28-607–25.

Poag, C. W. Roadblocks on the kill curve: Testing the Raup hypothesis. *Palaios* 12 (1997): 582–90.

Pope, K. O. Impact dust not the cause of the Cretaceous-Tertiary mass extinction. *Geology* 30 (2002): 99–102.

Pope, K. O., K. H. Baines, A. C. Ocampo, and B. A. Ivanov. Energy, volatile production, and climatic effects of the Chicxulub Cretaceous/Tertiary impact. *Journal of Geophysical Research* 102 (1997): 21-645–64.

Pope, K. O., K. H. Baines, A. C. Ocampo, and B. A. Ivanov. Impact winter and the Cretaceous/Tertiary extinctions: Results of a Chicxulub asteroid impact model. *Earth and Planetary Science Letters* 128 (1994): 719–25.

Prinn, R. G., and B. Fegley Jr. Bolide impacts, acid rain, and biospheric traumas at the Cretaceous-Tertiary boundary. *Earth and Planetary Science Letters* 83 (1987): 1–15.

Rampino, M. R. Role of the galaxy in periodic impacts and mass extinctions on the Earth. *Geological Society of America Special Papers* 356 (2002): 667–78.

Rampino, M. R., and K. Caldeira. Periodic impact cratering and extinction events over the last 260 million years. *Monthly Notices of the Royal Astronomical Society* 454 (2015): 3480–84.

Rampino, M. R., and B. M. Haggerty. Extraterrestrial impacts and mass extinctions of life. In *Hazards Due to Asteroids and Comets*, edited by T. Gehrels, 827–57. Tucson: University of Arizona Press, 1994.

Rampino, M. R., and T. Volk. Mass extinctions, atmospheric sulphur and climatic warming at the K/T boundary. *Nature* 332 (1988): 63–65.

Raup, D. M. Impact as a general cause of extinction: A feasibility test. *Geological Society of America Special Papers* 247 (1990): 27–32.

Raup, D. M. A kill curve for Phanerozoic marine species. *Paleobiology* 17 (1991): 37–48.

Shoemaker, E. M., and R. F. Wolfe. Mass extinctions, crater ages, and comet showers. In *The Galaxy and the Solar System*, edited by R. Smoluchowski, J. N. Bahcall, and M. S. Matthews, 338–86. Tucson: University of Arizona Press, 1986.

Shoemaker, E. M., R. F. Wolfe, and C. S. Shoemaker. Asteroid and comet flux in the neighborhood of Earth. *Geological Society of America Special Papers* 247 (1990): 155–70.

Steiner, M. B., and E. M. Shoemaker. An hypothesized Manson impact tsunami: Paleomagnetic and stratigraphic evidence in the Crow Creek Member, Pierre Shale. *Geological Society of America Special Papers* 302 (1996): 419–32.

Toon, O. B., K. Zahnle, D. Morrison, R. P. Turco, and C. Covey. Environmental perturbations caused by the impacts of asteroids and comets. *Reviews of Geophysics* 35 (1997): 41–78.

Toon, O. B., K. Zahnle, R. P. Turco, and C. Covey. Environmental perturbations caused by asteroid impacts. In *Hazards Due to Asteroids and Comets*, edited by T. Gehrels, 791–826. Tucson: University of Arizona Press, 1994.

Vajda, V., J. I. Raine, and C. J. Hollis. Indication of global deforestation at the Cretaceous-Tertiary boundary by New Zealand fern spike. *Science* 294 (2007): 1700–1702.

Varricchio, D. Taphonomy of Jack's Birthday Site, a diverse dinosaur bonebed from the Upper Cretaceous Two Medicine Formation of Montana. *Palaeogeography, Palaeoclimatology, Palaeoecology* 114 (1995): 297–323.

Varricchio, D. J., and J. R. Horner. Hadrosaurid and lambeosaurid bone beds from the Upper Cretaceous Two Medicine Formation of Montana: Taphonomic and biologic implications. *Canadian Journal of Earth Sciences* 30 (1993): 997–1006.

Vellekoop, J., S. Esmeray-Senlet, K. G. Miller, J. V. Browning, A. Sluijs, B. van de Schootbrugge, J. S. Sinninghe Damsté, and H. Brinkhuis. Evidence for Cretaceous-Paleogene boundary bolide "impact winter" conditions from New Jersey, USA. *Geology* 44 (2016): 619–22.

Vickery, A. M., and H. J. Melosh. Atmospheric erosion and impactor retention in large impacts, with application to mass extinctions. *Geological Society of America Special Papers* 247 (1990): 289–99.

Wang, W., and T. J. Ahrens. Shock vaporization of anhydrite and global effects of the K/T bolide. *Earth and Planetary Science Letters* 156 (1998): 125–40.

Wolbach, W. S., I. Gilmour, E. Anders, C. J. Orth, and R. R. Brooks. Global fire at the Cretaceous-Tertiary boundary. *Nature* 334 (1988): 670–73.

Wolfe, J. A. Palaeobotanical evidence for a marked temperature increase following the Cretaceous/Tertiary boundary. *Nature* 343 (1990): 153–56.

Wolfe, J. A., and G. R. Upchurch. Vegetation, climatic and floral changes at the Cretaceous-Tertiary boundary. *Nature* 324 (1986): 148–52.

Zachos, J. C., and M. A. Arthur. Paleoceanography of the Cretaceous/Tertiary boundary event: Inferences from stable isotopic and other data. *Paleoceanography* 1 (1986): 5–26.

Zachos, J. C., M. A. Arthur, and W. E. Dean. Geochemical evidence for suppression of pelagic marine productivity at the Cretaceous/Tertiary boundary. *Nature* 337 (1989): 61–64.

Zahnle, K. Atmospheric chemistry by large impacts. *Geological Society of America Special Papers* 247 (1990): 271–88.

6. Catastrophism and Natural Selection

Darwin, C. Natural selection. Letter to *Gardeners' Chronicle and Agricultural Gazette*, April 21, 1860, 362–63.

Darwin, C. *On the Origin of Species by Means of Natural Selection or the Preservation of Favored Races in the Struggle for Survival.* London: Murray, 1859.

Dempster, W. J. *Evolutionary Concepts of the Nineteenth Century: Natural Selection and Patrick Matthew.* Durham: Pentland Press, 1996.

Eiseley, L. *Darwin's Century: Evolution and the Men Who Discovered It.* New York: Doubleday, 1959.

Gould, S. J. *The Flamingo's Smile: Reflections in Natural History.* New York: Norton, 1985.

Lyell, C. *Principles of Geology, Being an Attempt to Explain the Former Changes of the Earth's Surface by Processes Still in Operation.* 3 vols. London: Murray, 1830–1833.

Matthew, P. Nature's law of selection. Letter to *Gardeners' Chronicle and Agricultural Gazette*, April 7, 1860, 312–13.

Matthew, P. *On Naval Timber and Arboriculture; with Critical Notes on Authors Who Have Recently Treated the Subject of Planting.* Edinburgh: Black, 1831.

Mayr, E. *Animal Species and Evolution.* Cambridge, Mass.: Harvard University Press, 1963.

Mayr, E. *Toward a New Philosophy of Biology.* Cambridge, Mass.: Harvard University Press, 1989.

McKinney, H. L. *Wallace and Natural Selection.* New Haven, Conn.: Yale University Press, 1972.

Rampino, M. R. Darwin's error? Patrick Matthew and the catastrophic nature of the geologic record. *Historical Biology* 23 (2011): 227–30.

Rudwick, M. J. S. *Georges Cuvier, Fossil Bones, and Geological Catastrophes: New Translations and Interpretations of the Primary Texts.* Chicago: University of Chicago Press, 1997.

Rudwick, M. J. S. *The Meaning of Fossils: Episodes in the History of Palaeontology.* 2nd ed. Chicago: University of Chicago Press, 1976.

Wainwright, M. Natural selection: It's not Darwin's (or Wallace's) theory. *Saudi Journal of Biological Sciences* 15 (2008): 1–8.

Wainwright, M. The origin of species without Darwin and Wallace. *Saudi Journal of Biological Sciences* 17 (2010): 187–204.

Wallace, A. R. *My Life: A Record of Events and Opinions.* Vol. 2. New York: Dodd, Meade, 1905.

7. Impacts and Extinctions

General

Alvarez, W., F. Asaro, and A. Montanari. Iridium profile for 10 million years across the Cretaceous-Tertiary boundary at Gubbio (Italy). *Science* 250 (1990): 1700–1702.

Clube, S. V. M., and W. M. Napier. The role of episodic bombardment in geophysics. *Earth and Planetary Science Letters* 57 (1982): 251–62.

Glass, B. P., and B. M. Simonson. *Distal Impact Ejecta Layers: A Record of Large Impacts in Sedimentary Deposits.* Berlin: Springer, 2013.

Gostin, V. A., R. R. Keays, and M. W. Wallace. Iridium anomaly from the Acraman impact ejecta horizon: Impacts can produce sedimentary iridium peaks. *Nature* 340 (1989): 542–44.

Hallam, A. The case for sea-level change as a dominant causal factor in mass extinction of marine invertebrates. *Philosophical Transactions of the Royal Society of London B* 325 (1989): 437–55.

Hallam, A. Major bio-events in the Triassic and Jurassic. In *Global Events and Event Stratigraphy in the Phanerozoic,* edited by O. H. Walliser, 265–83. Berlin: Springer, 1995.

Holser, W. T., and M. Magaritz. Cretaceous/Tertiary and Permian/Triassic boundary events compared. *Geochimica et Cosmochimica Acta* 56 (1992): 3297–309.

Holser, W. T., M. Magaritz, and R. L. Ripperdan. Global isotopic events. In *Global Events and Event Stratigraphy in the Phanerozoic,* edited by O. H. Walliser, 63–88. Berlin: Springer, 1995.

Hsü, K. J., and J. A. McKenzie. A Strangelove ocean in earliest Tertiary. In *Carbon Cycle and Atmospheric CO_2: Natural Variations Archean to Present,* edited by E. T. Sundquist and W. S. Broecker, 32:487–92. Washington, D.C.: American Geophysical Union, 1985.

Kasting, J. F., S. M. Richardson, J. B. Pollack, and O. B. Toon. A hybrid model of the CO_2 geochemical cycle and its application to large impact events. *American Journal of Science* 286 (1986): 361–89.

Kump, L. R. Interpreting carbon-isotope excursions: Strangelove oceans. *Geology* 19 (1991): 299–302.

Kump, L. R., and M. A. Arthur. Interpreting carbon-isotope excursions: Carbonates and organic matter. *Chemical Geology* 161 (1999): 181–98.

Kyte, F. T., and D. E. Brownlee. Unmelted meteoritic debris in a Late Pliocene iridium anomaly: Evidence for the ocean impact of a nonchondritic asteroid. *Geochimica et Cosmochimica Acta* 49 (1985): 1095–108.

Kyte, F. T., Z. Zhou, and J. T. Wasson. High noble metal concentrations in a late Pliocene sediment. *Nature* 292 (1981): 417–20.

Magaritz, M., W. T. Holser, and J. L. Kirschvink. Carbon-isotope events across the Precambrian/Cambrian boundary on the Siberian Platform. *Nature* 320 (1986): 258–59.

Margolis, S. V., P. Claeys, and F. T. Kyte. Microtektites, microkrystites, and spinels from a Late Pliocene asteroid impact in the Southern Ocean. *Science* 251 (1991): 1594–97.

McLaren, D. J., and W. D. Goodfellow. Geological and biological consequences of giant impacts. *Annual Review of Earth and Planetary Sciences* 18 (1990): 123–71.

Milne, D. H., and C. P. McKay. Response of marine plankton communities to a global atmospheric darkening. *Geological Society of America Special Papers* 190 (1982): 297–304.

Orth, C. J. Geochemistry of the bio-event horizons. In *Mass Extinctions: Processes and Evidence*, edited by S. K. Donovan, 37–72. New York: Columbia University Press, 1989.

Orth, C. J., M. Attrep Jr., and L. R. Quintana. Iridium abundance patterns across bio-event horizons in the fossil record. *Geological Society of America Special Papers* 247 (1990): 45–59.

Orth, C. J., M. Attrep Jr., L. R. Quintana, W. P. Elder, E. G. Kauffman, R. Diner, and T. Villamil. Elemental abundance anomalies in the late Cenomanian extinction interval: A search for the source(s). *Earth and Planetary Science Letters* 117 (1993): 189–204.

Rampino, M. R. Are marine and nonmarine extinctions correlated? *Eos, Transactions of the American Geophysical Union* 69 (1988): 889–95.

Rampino, M. R., B. M. Haggerty, and T. C. Pagano. A unified theory of impact crises and mass extinctions: Quantitative tests. *Annals of the New York Academy of Sciences* 822 (1997): 403–31.

Rampino, M. R., and T. Volk. Mass extinctions, atmospheric sulphur and climatic warming at the K/T boundary. *Nature* 332 (1988): 63–65.

Rigby, J. K., Jr., and D. L. Wolberg. The Therian mammalian fauna (Campanian) of Quarry 1, Fossil Forest study area, San Juan Basin, New Mexico. *Geological Society of America Special Papers* 209 (1987):51–79.

Roccia, R., D. Boclet, P. Bonte, A. Castellarin, and C. Jehanno. An iridium anomaly in the Middle-Lower Jurassic of the Venetian region, northern Italy. *Journal of Geophysical Research* 91 (1986): E259–62.

Rogers, R. R. Taphonomy of three dinosaur bonebeds in the Upper Cretaceous Two Medicine Formation of Northwestern Montana: Evidence for drought-related mortality. *Palaios* 5 (1990): 394–413.

The Late Eocene

Alvarez, W., L. W. Alvarez, F. Asaro, and H. V. Michel. Iridium anomaly approximately synchronous with terminal Eocene extinctions. *Science* 216 (1982): 886–88.

Bodiselitisch, B., A. Montanari, C. Koeberl, and R. Coccioni. Delayed climate cooling in the Late Eocene caused by multiple impacts: High-resolution geochemical studies at Massignano, Italy. *Earth and Planetary Science Letters* 233 (2004): 283–302.

Bottomley, R., R. Grieve, and V. Massaitis. The age of the Popigai impact event and its relationship to events at the Eocene/Oligocene boundary. *Nature* 388 (1997): 365–68.

Clymer, A., D. Bice, and A. Montanari. Shocked quartz from the late Eocene: Impact evidence from Massignano, Italy. *Geology* 24 (1996): 483–86.

Coccioni, R., D. Basso, H. Brinkhuis, S. Galeotti, S. Gardin, S. Monechi, and S. Spezzaferri. Marine biotic signals across a late Eocene impact layer at Massignano, Italy: Evidence for long-term environmental perturbations? *Terra Nova* 6 (2000): 258–63.

Coccioni, R., F. Fontalini, and S. Spezzaferri. Late Eocene impact-induced climate and hydrological changes: Evidence from the Massignano global stratotype section and point (Central Italy). *Geological Society of America Special Papers* 452 (2009): 97–118.

Farley, K. Cenozoic variations in the flux of interplanetary dust recorded by ^{3}He in deep-sea sediment. *Nature* 376 (1995): 153–56.

Farley, K. A., A. Montanari, E. M. Shoemaker, and C. S. Shoemaker. Geochemical evidence for a comet shower in the Late Eocene. *Science* 280 (1998): 1250–53.

Ganapathy, R. Evidence for a major meteorite impact on the Earth 34 million years ago: Implications for Eocene extinctions. *Science* 216 (1982): 885–86.

Glass, B. P. Possible correlations between tektite events and climatic changes? *Geological Society of America Special Papers* 190 (1982): 251–56.

Glass, B. P., R. N. Baker, D. Strozer, and G. A. Wagner. North American microtektites from the Caribbean Sea and their fission-track ages. *Earth and Planetary Science Letters* 19 (1973): 184–92.

Glass, B. P., D. L. DuBois, and R. Ganapathy. Relationship between an iridium anomaly and the North American microtektite layer in core RC9-58 from the Caribbean Sea. *Journal of Geophysical Research* 87, suppl. (1982): A425–28.

Glass, B. P., and B. M. Simonson. *Distal Impact Ejecta Layers: A Record of Large Impacts in Sedimentary Deposits.* Berlin: Springer, 2013.

Gohn, G. S., C. Koeberl, K. G. Miller, W. U. Reimold, J. V. Browning, C. S. Cockell, J. W. Horton Jr., T. Kenkmann, A. A. Kulpecz, D. S. Powars, et al. Deep drilling into the Chesapeake Bay impact structure. *Science* 320 (2008): 1740–45.

Grieve, R. A. F. Chesapeake Bay and other terminal Eocene impacts. *Meteoritics & Planetary Science* 31 (1996): 166–67.

Keller, G. Stepwise mass extinctions and impact events: Late Eocene to early Oligocene. *Marine Micropaleontology* 10 (1986): 267–93.

Koeberl, C., C. W. Poag, W. U. Reimold, and D. Brandt. Impact origin of the Chesapeake Bay structure and the source of the North American tektites. *Science* 271 (1996): 1263–66.

Masaitis, V. L., M. V. Naumov, and M. S. Mashchak. Anatomy of the Popigai impact crater, Russia. *Geological Society of America Special Papers* 339 (1999): 1–17.

Miller, K. G., W. A. Berggren, J. Zhang, and A. A. Palmer-Julson. Biostratigraphy and isotope stratigraphy of Upper Eocene microtektites at Site 612: How many impacts? *Palaios* 6 (1991): 17–38.

Monechi, S., A. Buccianti, and S. Gardin. Biotic signals from nannoflora across the iridium anomaly in the upper Eocene of the Massignano section: Evidence from statistical analysis. *Marine Micropaleontology* 39 (2000): 219–37.

Montanari, A. Geochronology of the terminal Eocene impacts: An update. *Geological Society of America Special Papers* 247 (1990): 607–16.

Montanari, A., F. Asaro, H. V. Michel, and J. P. Kennett. Iridium anomalies of Late Eocene age at Massignano (Italy), and ODP Site 689B (Maud Rise, Antarctica). *Palaios* 8 (1993): 420–37.

Pierrard, O., E. Robin, R. Rocchia, and A. Montanari. Extraterrestrial Ni-rich spinel in upper Eocene sediments from Massignano, Italy. *Geology* 26 (1998): 307–10.

Poag, C. W. The Chesapeake Bay bolide impact: A convulsive event in Atlantic Coastal Plain evolution. *Sedimentary Geology* 108 (1997): 45–90.

Poag, C. W., and L. J. Poppe. The Toms Canyon structure, New Jersey outer continental shelf: A possible late Eocene impact crater. *Marine Geology* 145 (1998): 23–60.

Poag, C. W., D. S. Powars, L. J. Poppe, and R. B. Mixon. Meteoroid mayhem in Ole Virginny: Source of the North American tektite strewn field. *Geology* 22 (1994): 691–94.

Poag, C. W., D. S. Powars, L. J. Poppe, R. B. Mixon, L. E. Edwards, D. W. Folger, and S. Bruce. Deep Sea Drilling Project Site 612 bolide event: New evidence of a late Eocene impact-wave deposit and possible impact site, U.S. east coast. *Geology* 20 (1992): 771–74.

Prothero, D. R. *The Eocene-Oligocene Transition: Paradise Lost.* New York: Columbia University Press, 1994.

Sanfilippo, A., W. R. Riedel, B. P. Glass, and F. T. Kyte. Late Eocene microtektites and radiolarian extinctions on Barbados. *Nature* 314 (1985): 613–15.

Schmitz, B., S. Boschi, A. Cronholm, P. R. Heck, S. Monechi, A. Montanari, and F. Terfelt. Fragments of Late Eocene Earth-impacting asteroids linked to disturbance of asteroid belt. *Earth and Planetary Science Letters* 425 (2015): 77–83.

Vishnevsky, S., and A. Montanari. Popigai impact structure (Arctic Siberia, Russia): Geology, petrology, geochemistry, and geochronology of glass-bearing impactites. *Geological Society of America Special Papers* 339 (1999): 19–59.

The Jurassic/Cretaceous Boundary

Corner, B., W. U. Reimold, D. Bandt, and C. Koeberl. Morokweng impact structure, Northwest Province, South Africa: Geophysical imaging and shock petrographic studies. *Earth and Planetary Science Letters* 146 (1997): 351–64.

Deconinck, J. F., F. Baudin, and N. Tribovillard. The Purbeckian facies of the Boulonnais: A tsunami deposit hypothesis (Jurassic-Cretaceous boundary, northern France). *Comptes Rendus de l'Académie de sciences: Earth and Planetary Science* 330 (2000): 527–32.

Dypvik, H., S. T. Gudlaugsson, F. Tsikalas, M. Attrep Jr., R. E. Ferrell Jr., D. H. Krinsley, A. Mørk, J. I. Faleide, and J. Nagy. Mjølnir structure: An impact crater in the Barents Sea. *Geology* 24 (1996): 779–82.

Hart, R. J., M. A. G. Andreoli, M. Tredoux, D. Moser, L. D. Ashwal, E. A. Eide, S. J. Webb, and D. Brandt. Late Jurassic age for the Morokweng impact structure, Southern Africa. *Earth and Planetary Science Letters* 147 (1997): 25–35.

Houša, V., P. Pruner, V. A. Zakharov, M. Košt'ák, M. Chadima, M. A. Rogov, S. Slechta, and M. Mazuch. Boreal-Tethyan correlation of the Jurassic-Cretaceous boundary interval by magneto- and biostratigraphy. *Stratigraphy and Geological Correlation* 15 (2007): 297–309.

Irvine, G. J., I. McDonald, A. S. Gale, and W. U. Reimold. Platinum-group elements from the Purbeck Cinder Bed of England and the Boulonnais of France: Implications for an impact event at the Jurassic-Cretaceous boundary. *Meteoritics & Planetary Science* 38 (2003): 2–23.

Kudielka, G., C. Koeberl, A. Montanari, J. Newton, and W. U. Reimold. Stable-isotope and trace-element stratigraphy of the Jurassic/Cretaceous boundary, Bosso River Gorge, Italy. In *Geological and Biological Effects of Impact Events*, edited by E. Buffetaut and C. Koeberl, 25–68. Berlin: Springer, 2000.

Kudielka, G., C. Koeberl, A. Montanari, J. Newton, and W. U. Reimold. Stable isotope stratigraphy of the J-K boundary Bosso Gorge, Italy. *Geochimica et Cosmochimica Acta* 58 (1999): 1393–97.

McDonald, I., G. J. Irvine, E. de Vos, A. S. Gale, and W. U. Reimold. Geochemical search for impact signatures in possible impact-generated units associated with the Jurassic-Cretaceous boundary in southern England and northern France. In *Biological Processes Associated with Impact Events*, edited by C. Cockell, I. Gilmour, and C. Koeberl, 257–86. Berlin: Springer, 2006.

Reimold, W. U., R. A. Armstrong, and C. Koeberl. A deep drillcore from the Morokweng impact structure, South Africa: Petrography, geochemistry, and constraints on the crater size. *Earth and Planetary Science Letters* 201 (2002): 221–32.

Smelror, M., S. R. A. Kelly, H. Dypvik, A. Mørk, J. Nagy, and F. Tsikalas. Mjølnir (Barents Sea) meteorite impact ejecta offers a Volgian-Ryazanian boundary marker. *Newsletters on Stratigraphy* 38 (2001): 129–40.

Tremolada, F., A. Bornemann, T. J. Bralower, C. Koeberl, and B. van de Schootbrugge. Paleoceanographic changes across the Jurassic/Cretaceous boundary: The calcareous phytoplankton response. *Earth and Planetary Science Letters* 241 (2006): 361–71.

Wimbledon, A. A. P. The Jurassic-Cretaceous boundary: An age-old correlative enigma. *Episodes* 31 (2008): 423–28.

Zakharov, V. A., A. S. Lapukhov, and O. V. Shenfilk. Iridium anomaly at the Jurassic-Cretaceous boundary in northern Siberia. *Russian Journal of Geology and Geophysics* 34 (1993): 83–90.

The Bajocian/Bathonian Boundary

Jehanno, C., D. Boclet, P. Bonté, A. Castellarin, and R. Rocchia 1988. Identification of two populations of extraterrestrial particles in a Jurassic hardground of the Southern Alps. In *Proceedings of the 18th Lunar and Planetary Science Conference*, edited by G. Ryder, 623–30. Cambridge: Cambridge University Press, 1988.

Pallfy, J. Did the Puchezh-Katunki impact trigger an extinction? In *Cratering in Marine Environments and on Ice*, edited by H. Dypvik, M. Burchell, and P. Claeys, 135–48. Berlin: Springer, 2004.

Rocchia, R., D. Boclet, P. Bonté, A. Castellarin, and C. Jehanno. An iridium anomaly in the Middle-Lower Jurassic of the Venetian Region, Northern Italy. *Journal of Geophysical Research* 91 (1986): E259–62.

The Middle Norian

Kirkham, A. Glauconitic spherules from the Triassic of the Bristol area, SW England: Probable mictotektite pseudomorph. *Proceedings of the Geologists' Association* 114 (2003): 11–21.

Onoue, T., H. Sato, T. Nakamura, T. Noguchi, Y. Hidaka, N. Shirai, M. Ebihara et al. Deep-sea record of impact apparently unrelated to mass extinction in the Late Triassic. *Proceedings of the National Academy of Sciences of the United States of America* 109 (2012): 19134–39.

Parker, W. G., and J. W. Martz. The Late Triassic (Norian) Adamanian-Revueltian tetrapod faunal transition in the Chinle Formation of Petrified Forest National Park. *Earth and Environmental Science Transactions of the Royal Society of Edinburgh* 101 (2010): 231–60.

Sato, H., T. Onoue, T. Nozaki, and K. Suzuki. Osmium isotope evidence for a large Late Triassic impact event. *Nature Communications* 4 (2013):1–7.

Thackrey, S., G. Walkden, A. Indares, M. Horstwood, S. Kelley, and R. Parrish. The use of heavy mineral correlation for determining the source of impact ejecta: A Manicouagan distal ejecta case study. *Earth and Planetary Science Letters* 285 (2009): 163–72.

Walkden, G., J. Parker, and S. Kelley. A Late Triassic impact ejecta layer in south-western Britain. *Science* 298 (2013): 2185–88.

Wolfe, S. H. Potassium-argon ages of the Manicouagan-Mushalagan Lakes Structure. *Journal of Geophysical Research* 76 (1971): 5424–36.

The Late Devonian

Bai, S. L., Z. Q. Bai, X. P. Ma, D. R. Wang, and Y. L. Sun. *Devonian Events and Biostratigraphy of South China: Conodont Zonation and Correlation, Bio-Event and Chemo-Event, Milankovitch Cycle and Nickel-Episode*. Beijing: Beijing University Press, 1994.

Bond, D., and P. B. Wignall. Evidence for Late Devonian (Kellwasser) anoxic events in the Great Basin, western United States. In *Understanding Late Devonian and Permian-Triassic Biotic and Climatic Events: Towards an Integrated Approach*, edited by D. J. Over, J. R. Morrow, and P. B. Wignal, 225–62. Amsterdam: Elsevier, 2005.

Bond, D., P. B. Wignall, and G. Racki. Extent and duration of marine anoxia during the Frasnian-Famennian (Late Devonian) mass extinction in Poland, Germany, Austria and France. *Geological Magazine* 141 (2004): 173–93.

Bridge, J. S., and M. L. Droser. Unusual marginal-marine lithofacies from the Upper Devonian Catskill clastic wedge. *Geological Society of America Special Papers* 201 (1985):143–61.

Casier, J.-G., and F. Lethiers. Ostracods and the late Devonian mass extinction: The Schmidt quarry parastratotype (Kellerwald, Germany). *Comptes Rendus de l'Académie de sciences: Earth and Planetary Science* 326 (1998): 71–78.

Casier, J.-G., F. Lethiers, and P. Claeys. Ostracod evidence for an abrupt mass extinction at the Frasnian/Famennian boundary (Devils Gate, Nevada, USA). *Comptes Rendus de l'Académie de sciences: Earth and Planetary Science* 322 (1996): 415–22.

Chai, Z.-F., X.-Y. Mao, S.-L. Ma, S.-L. Bai, C. J. Orth, Y.-Q. Zhou, and J.-G. Ma. Geochemical anomaly of the Devonian-Carboniferous boundary at Huangmao, Guangxi, China. *Abstracts, International Geological Correlation Project 199, Rare Events in Geology* (1987): 29.

Chen, D., and M. E. Tucker. The Frasnian-Famennian mass extinction: Insights from high-resolution sequence stratigraphy and cyclostratigraphy in South China. *Palaeogeography, Palaeoclimatology, Palaeoecology* 193 (2003): 87–111.

Chen, D., M. E. Tucker, Y. Shen, J. Yans, and A. Preat. Carbon isotope excursions and sea-level change: Implications for the Frasnian-Famennian biotic crisis. *Journal of the Geological Society* 159 (2002): 623–26.

Claeys, P., and J.-G. Casier. Microtektite-like impact glass associated with the Frasnian-Famennian boundary extinction. *Earth and Planetary Science Letters* 122 (1994): 303–18.

Claeys, P., J.-G. Casier, and S. V. Margolis. Microtektites and mass extinctions: Evidence for a Late Devonian asteroid impact. *Science* 257 (1992): 1102–4.

Claeys, P., F. T. Kyte, A. Herbosch, and J.-G. Casier. Geochemistry of the Frasnian-Famennian boundary, Belgium: Mass extinction, anoxic oceans and microtektite layer, but not much iridium? *Geological Society of America Special Papers* 307 (1996): 491–504.

Ellwood, B. B., S. L. Benoist, A. El Hassani, C. Wheeler, and R. E. Crick. Impact ejecta layer from the mid-Devonian: Possible connection to global mass extinctions. *Science* 300 (2003): 1734–37.

Girard, C., E. Robin, R. Rocchia, L. Froget, and R. Feist. Search for impact remains at the Frasnian-Famennian boundary in the stratotype area, southern France. *Palaeogeography, Palaeoclimatology, Palaeoecology* 132 (1997): 391–97.

Gong, Y.-M., B.-H. Li, C.-Y. Wang, and Y. Wu. Orbital cyclostratigraphy of the Devonian Frasnian-Famennian transition in China. *Palaeogeography, Palaeoclimatology, Palaeoecology* 168 (2001): 237–48.

Goodfellow, W. D., H. H. J. Geldsetzer, D. J. McLaren, M. J. Orchard, and G. Klapper. Geochemical and isotopic anomalies associated with the Frasnian-Famennian extinction. *Historical Biology* 2 (1989): 51–72.

Joachimski, M. M. Comparison of organic and inorganic carbon isotope patterns across the Frasnian-Famennian boundary. *Palaeogeography, Palaeoclimatology, Palaeoecology* 132 (1997): 133–45.

Ma, X. P., and S. L. Bai. Biological, depositional, microspherule, and geochemical records of the Frasnian/Famennian boundary beds, South China. *Palaeogeography, Palaeoclimatology, Palaeoecology* 181 (2002): 325–46.

McGhee, G. R., Jr. The Frasnian-Famennian extinction event. In *Mass Extinctions: Processes and Evidence*, edited by S. K. Donovan, 133–51. New York: Columbia University Press, 1989.

McGhee, G. R., Jr. The Late Devonian extinction event: Evidence for abrupt ecosystem collapse. *Paleobiology* 14 (1988): 250–57.

McGhee, G. R., Jr. *The Late Devonian Mass Extinction*. New York: Columbia University Press, 1996.

McGhee, G. R., Jr. The multiple impacts hypothesis for mass extinction: A comparison of the Late Devonian and the late Eocene. *Palaeogeography, Palaeoclimatology, Palaeoecology* 176 (2001): 47–58.

McGhee, G. R., Jr. *When Invasion of the Land Failed: The Legacy of the Devonian Extinctions*. New York: Columbia University Press, 2013.

McLaren, D. J. Mass extinction and iridium anomaly in the Upper Devonian of Western Australia: A commentary. *Geology* 13 (1985): 170–72.

Nicoli, R. S., and P. E. Playford. Upper Devonian iridium anomalies, conodont zonation and the Frasnian-Famennian boundary in the Canning Basin, Western Australia. *Palaeogeography, Palaeoclimatology, Palaeoecology* 104 (1993): 105–13.

Playford, P. E., D. J. McLaren, C. J. Orth, J. S. Gilmore, and W. D. Goodfellow. Iridium anomaly in the Upper Devonian of the Canning Basin, Western Australia. *Science* 226 (1984): 437–39.

Racki, G. The Frasnian-Famennian biotic crisis: How many (if any) bolide impacts? *Geologische Rundschau* 87 (1999): 617–32.

Reimold, W. U., S. P. Kelley, S. C. Sherlock, H. Henkel, and C. Koeberl. Laser argon dating of melt breccias from the Siljan impact structure, Sweden: Implication for a possible relationship to Late Devonian extinction events. *Meteoritics & Planetary Science* 40 (2005): 591–607.

Schindler, E. Event-stratigraphic markers within the Kellwasser Crisis near the Frasnian-Famennian boundary (Upper Devonian) in Germany. *Palaeogeography, Palaeoclimatology, Palaeoecology* 104 (1993): 115–25.

Wang, K. Glassy microsphrules (microtektites) from an Upper Devonian limestone. *Science* 256 (1992): 1547–50.

Wang, K., M. Attrep Jr., and C. J. Orth. Global iridium anomaly, mass extinction, and redox change at the Devonian-Carboniferous boundary. *Geology* 21 (1993): 1071–74.

Wang, K., and H. H. J. Geldsetzer. Late Devonian conodonts define the precise horizon of the Frasnian-Famennian boundary at Cinquefoil Mountain, Jasper, Alberta. *Canadian Journal of Earth Sciences* 32 (1995): 1825–34.

Wang, K., H. H. J. Geldsetzer, and B. D. E. Chatterton. A Late Devonian extraterrestrial impact and extinction in eastern Gondwana: Geochemical, sedimentological, and faunal evidence. *Geological Society of America Special Papers* 293 (1994): 111–20.

Wang, K., H. H. J. Geldsetzer, W. D. Goodfellow, and H. R. Krouse. Carbon and sulfur isotope anomalies across the Frasnian-Famennian extinction boundary, Alberta, Canada. *Geology* 24 (1996): 187–91.

Wang, K., C. J. Orth, M. Attrep Jr., B. D. E. Chatterton, H. Hou, and H. H. J. Geldsetzer. Geochemical evidence for a catastrophic biotic event at the Frasnian/Famennian boundary in south China. *Geology* 19 (1991): 776–79.

Warme, J. E., and H.-C. Kuehner. Anatomy of an anomaly: The Devonian catastrophic Alamo impact breccia of southern Nevada. *International Geology Review* 40 (1996): 189–216.

Warme, J. E., and C. A. Sandberg. Alamo megabreccia: Record of Late Devonian impact in southern Nevada. *GSA Today* 6 (1996): 1–7.

8. The Great Dying

Becker, L., R. J. Poreda, A. R. Basu, K. O. Pope, T. M. Harrison, C. Nicholson, and R. Iasky. Bedout: A possible end-Permian crater offshore of northwestern Australia. *Science* 304 (2004): 1469–76.

Becker, L., R. J. Poreda, A. G. Hunt, T. E. Bunch, and M. Rampino. Impact event at the Permian-Triassic boundary: Evidence from extraterrestrial noble gases in fullerenes. *Science* 291 (2001): 1530–33.

Bercovici, A., Y. Cui, M.-B. Forel, J. Yu, and V. Vajda. Terrestrial paleoenvironment characterization across the Permian-Triassic boundary in South China. *Journal of Asian Earth Sciences* 98 (2015): 225–46.

Bercovici, A., and V. Vajda. Terrestrial Permian-Triassic boundary sections in South China. *Global and Planetary Change* 143 (2016): 312–33.

Bowring, S. A., D. H. Erwin, Y. G. Jin, M. W. Martin, K. Davidek, and W. Wang. U/Pb zircon geochronology and tempo of the end-Permian mass extinction. *Science* 280 (1998): 1039–45.

Burgess, S. D., S. Bowring, and S.-Z. Shen. High-precision timeline for Earth's most severe extinction. *Proceedings of the National Academy of Sciences of the United States of America* 111 (2014): 3316–21.

Cao, C., G. D. Love, L. E. Hays, W. Wang, S. Shen, and R. E. Summons. Biogeochemical evidence for euxinic oceans and ecological disturbance presaging the end-Permian mass extinction event. *Earth and Planetary Science Letters* 81 (2009): 188–201.

Carrasquillo, A. J., C. Cao, D. H. Erwin, and R. E. Summons. Non-detection of C_{60} fullerene at two mass extinction horizons. *Geochimica et Cosmochimica Acta* 176 (2016): 18–25.

Chen, Z.-Q., J. Tong, K. Kaiho, and H. Kawahata. Onset of biotic and environmental recovery from the end-Permian mass extinction within 1–2 million years: A case study of the Lower Triassic of the Meishan section, South China. *Palaeogeography, Palaeoclimatology, Palaeoecology* 252 (2007): 176–87.

Chen, Z.-Q., H. Yang, M. Luo, M. J. Benton, K. Kaiho, L. Zhao, Y. Huang, K. Zhang, Y. Fang, H. Jiang, et al. Complete biotic and sedimentary records of the Permian-Triassic transition from Meishan section, South China: Ecologically assessing mass extinction and its aftermath. *Earth-Science Reviews* 149 (2015): 67–107.

Cirilli, S., C. P. Radrizzani, M. Ponton, and S. Radrizzani. Stratigraphical and palaeoenvironmental analysis of the Permian-Triassic transition in the Badia Valley (Southern Alps, Italy). *Palaeogeography, Palaeoclimatology, Palaeoecology* 138 (1998): 85–113.

Clarkson, M. O., S. A. Kasemann, R. A. Wood, T. M. Lentor, S. J. Daines, S. Richoz, F. Ohnemueller, A. Meixner, S. W. Poulton, and E. T. Tipper. Ocean acidification and the Permo-Triassic mass extinction. *Science* 348 (2015): 229–32.

Cui, Y., and L. R. Kump. Global warming and the end-Permian extinction event: Proxy and modeling perspectives. *Earth-Science Reviews* 149 (2015): 5–22.

Dao-Yi, X., and Y. Zheng. Carbon isotope and iridium event markers near the Permian/Triassic boundary in the Meishan section, Zhejiang Province, China. *Palaeogeography, Palaeoclimatology, Palaeoecology* 104 (1993): 171–76.

Erwin, D. H. *Extinction: How Life on Earth Nearly Ended 250 Million Years Ago*. Princeton, N.J.: Princeton University Press, 2006.

Erwin, D. H. *The Great Paleozoic Crisis*. New York: Columbia University Press, 1993.

Eshet, Y., M. R. Rampino, and H. Visscher. Fungal event and palynological record of ecological crisis and recovery across the Permian-Triassic boundary. *Geology* 23 (1995): 967–70.

Farley,K.A., P. Ward, G. Garrison, and S. Mukhopadhyay, Absence of extraterrestrial ^{3}He in Permian-Triassic age sedimentary rocks. *Earth and Planetary Science Letters* 240 (2005): 265–75.

Geldsetzer, H. H. J., and H. R. Krouse. Permian-Triassic extinction: Organic δ^{13}C evidence from British Columbia, Canada. *Geology* 22 (1994): 580–84.

Grasby, S. E., B. Beauchamp, D. P. G. Bond, P. Wignall, C. Talavera, J. M. Galloway, K. Piepjohn, L. Reinhardt, and D. Blomeier. Progressive environmental deterioration in northwestern Pangea leading to the latest Permian extinction. *Geological Society of America Bulletin* 127 (2015): 1331–47.

Holser, W. T., H. P. Schoenlaub, K. Boeckelmann, and M. Magaritz. The Permian-Triassic of the Gartnerkofel-1 core (Carnic Alps, Austria): Synthesis and conclusions. *Abhandlungen der Geologischen Bundesanstalt* 45 (1991): 213–32.

Isozaki, Y., N. Shimizu, J. Yao, Z. Ji, and T. Matsuda. End-Permian extinction and volcanic-induced environmental stress: The Permian-Triassic boundary interval of lower-slope facies at Chaotian, South China. *Palaeogeography, Palaeoclimatology, Palaeoecology* 252 (2007): 218–38.

Jin, Y. G., Y. Wang, W. Wang, Q. H. Shang, C. Q. Cao, and D. H. Erwin. Pattern of marine mass extinction near the Permian-Triassic boundary in South China. *Science* 289 (2000): 432–36.

Joachimski, M. M., X. Lai, S. Shen, H. Jiang, G. Luo, B. Chen, J. Chen, and Y. Sun. Climate warming in the latest Permian and the Permian-Triassic mass extinction. *Geology* 40 (2012): 195–98.

Kaiho, K., Z.-Q. Chen, H. Kawahata, Y. Kajiwara, and H. Sato. Close-up of the end-Permian mass extinction horizon recorded in the Meishan section, South China: Sedimentary, elemental, and biotic characterization and a negative shift in sulfate isotope ratio. *Palaeogeography, Palaeoclimatology, Palaeoecology* 239 (2006): 394–405.

Kershaw, S., T. Zhang, and G. Lan. A microbialite carbonate crust at the Permian-Triassic boundary in South China, and its palaeoenvironmental significance. *Palaeogeography, Palaeoclimatology, Palaeoecology* 146 (1999): 1–18.

Knoll, A. H., R. K. Bambach, D. E. Canfield, and J. P. Grotzinger. Comparative Earth history and Late Permian mass extinction. *Science* 273 (1996): 452–57.

Korte, C., and H. Kozur. Carbon-isotope stratigraphy across the Permian-Triassic boundary: A review. *Journal of Asian Earth Sciences* 39 (2010): 215–35.

Korte, C., P. Pande, P. Kalia, H. W. Kozur, M. M. Joachimski, and H. Oberhänsli. Massive volcanism at the Permian-Triassic boundary and its impact on the isotopic composition of the ocean and atmosphere. *Journal of Asian Earth Sciences* 37 (2010): 293–311.

Krull, E. S., and G. J. Retallack. Delta C-13 depth profiles from paleosols across the Permian-Triassic boundary: Evidence for methane release. *Geological Society of America Bulletin* 112 (2000): 1459–72.

Li, F., J. Yan, Z.-Q. Chen, J. G. Ogg, L. Tian, D. Korngreen, K. Liu, Z. Ma, and A. D. Woods. Global oolite deposits across the Permian-Triassic boundary: A synthesis and implications for palaeoceanography immediately after the end-Permian biocrisis. *Earth-Science Reviews* 149 (2015): 163–80.

Looy, C. V., W. A. Brugman, D. L. Dilcher, and H. Visscher. The delayed resurgence of equatorial forests after the Permian-Triassic ecologic crisis. *Proceedings of the National Academy of Sciences of the United States of America* 96 (1999): 13857–62.

Looy, C. V., R. J. Twitchett, D. L. Dilcher, J. H. A. van Konijnenburg-van Cittert, and H. Visscher. Life in the end-Permian dead zone. *Proceedings of the National Academy of Sciences of the United States of America* 98 (2001): 7879–83.

MacLeod, K. G., R. M. H. Smith, P. L. Koch, and P. D. Ward. Timing of mammal-like reptile extinctions across the Permian-Triassic boundary in South Africa. *Geology* 28 (2000): 227–30.

Magaritz, M. [13]C Minima follow extinction events: A clue to faunal radiation. *Geology* 17 (1989): 337–40.

Magaritz, M., and W. T. Holser. The Permian-Triassic of the Gartnerkofel-1 core (Carnic Alps, Austria): Carbon and oxygen isotope variation. *Abhandlungen der Geologischen Bundesanstalt* 45 (1991): 149–63.

Magaritz, M., R. V. Krishnamurthy, and W. T. Holser. Parallel trends in organic and inorganic carbon isotopes across the Permian/Triassic boundary. *American Journal of Science* 292 (1992): 727–39.

Michaelsen, P. Mass extinction of peat-forming plants and the effect on fluvial styles across the Permian-Triassic boundary, northern Bowen Basin, Australia. *Palaeogeography, Palaeoclimatology, Palaeoecology* 179 (2002): 173–88.

Morante, R., J. J. Veevers, A. S. Andrew, and P. J. Hamilton. Determination of the Permian-Triassic boundary in Australia from carbon isotope stratigraphy. *Australian Petroleum Exploration Association Journal* 34 (1994): 330–36.

Pang, Y., and G. R. Shi. Life crises on land across the Permian-Triassic boundary in South China. *Global and Planetary Change* 63 (2009): 155–65.

Payne, J. L., and M. E. Clapham. End-Permian mass extinction in the oceans: An ancient analog for the twenty-first century? *Annual Review of Earth and Planetary Science* 40 (2012): 89–111.

Payne, J. L., A. V. Turchyn, A. Paytan, D. J. DePaolo, D. J. Lehrmann, M. Yu, and J. Wei. Calcium isotope constraints on the end-Permian mass extinction. *Proceedings of the National Academy of Sciences of the United States of America* 107 (2010): 8543–48.

Pilkington, M., and R. A. F. Grieve. The geophysical signature of terrestrial impact craters. *Reviews of Geophysics* 30 (1992): 161–81.

Rampino, M. R., and A. C. Adler. Evidence for abrupt latest Permian mass extinction of foraminifera: Results of tests for the Signor-Lipps effect. *Geology* 26 (1998): 415–18.

Rampino, M. R., A. Prokoph, and A. C. Adler. Tempo of the end-Permian event: High-resolution cyclostratigraphy at the Permian-Triassic boundary. *Geology* 28 (2000): 643–46.

Reinhardt, J. W. Uppermost Permian reefs and Permo-Triassic sedimentary facies from the southeastern margin of Sichuan Basin, China. *Facies* 18 (1988): 231–88.

Retallack, G. J. Permian-Triassic life crisis on land. *Science* 267 (1995): 77–80.

Retallack, G. J. A 300-million-year record of atmospheric carbon dioxide from fossil plant cuticles. *Science* 411 (2001): 287–90.

Retallack, G. J., J. J. Veevers, and R. Morante. Global coal gap between Permian-Triassic extinction and Middle Triassic recovery of peat-forming plants. *Geological Society of America Bulletin* 108 (1996): 195–207.

Rocca, M. C. L., and J. L. B. Presser. A possible new very large impact structure in Malvinas Islands. *Historia Natural* 5 (2015): 121–33.

Rothman, D. H., G. P. Fournier, K. L. French, E. J. Alm, E. A. Boyle, C. Cao, and R. E. Summons. Methanogenic burst in the end-Permian carbon cycle. *Proceedings of the National Academy of Sciences of the United States of America* 111 (2014): 5462–67.

Sandler, A., Y. Eshet, and B. Schilman. Evidence for a fungal event, methane-hydrate release and soil erosion at the Permian-Triassic boundary in southern Israel. *Palaeogeography, Palaeoclimatology, Palaeoecology* 242 (2006): 68–89.

Sano, H., and K. Nakashima. Lowermost Triassic (Griesbachian) microbial bindstone-cementstone facies, southwest Japan. *Facies* 36 (1997): 1–24.

Schneebeli-Herman, E., W. M. Kürschner, P. A. Hochuli, D. Ware, H. Weissert, S. M. Bernasconi, G. Roohi, K. ur-Rehman, N. Goudemand, and H. Bucher. Evidence for atmospheric carbon injection during the end-Permian extinction. *Geology* 41 (2013): 579–82.

Shao, L., P. Zhang, J. Dou, and S. Shen. Carbon isotope compositions of the Late Permian carbonate rocks in southern China: Their variations between the Wujiaping and Changxing formations. *Palaeogeography, Palaeoclimatology, Palaeoecology* 161 (2000): 179–92.

Shen, J., T. J. Algeo, Q. Hu, G. Xu, L. Zhou, and Q. Feng. Volcanism in South China during the Late Permian and its relationship to marine ecosystem and environmental changes. *Global and Planetary Change* 105 (2013): 121–34.

Shen, J., Q. Feng, T. J. Algeo, C. Li, N. J. Planavsky, L. Zhou, and M. Zhang. Two pulses of oceanic environmental disturbance during the Permian-Triassic boundary crisis. *Earth and Planetary Science Letters* 443 (2016): 139–52.

Shen, W., Y. Lin, L. Xu, J. Li, Y. Wu, and Y. Sun. Pyrite framboids in the Permian-Triassic boundary section at Meishan, China: Evidence for dysoxic deposition. *Palaeogeography, Palaeoclimatology, Palaeoecology* 253 (2007): 323–31.

Smith, R. M. H., and P. D. Ward. Pattern of vertebrate extinctions across an event bed at the Permian-Triassic boundary in the Karoo Basin of South Africa. *Geology* 29 (2001): 1147–50.

Song, H., P. B. Wignall, D. Chu, J. Tong, Y. Sun, H. Song, W. He, and L. Tian. Anoxia/ high temperature double whammy during the Permian-Triassic marine crisis and its aftermath. *Scientific Reports* 4 (2014). doi: 10.1038/srep041342.

Song, H., P. B. Wignall, J. Tong, and H. Yin. Two pulses of extinction during the Permian-Triassic crisis. *Nature Geoscience* 6 (2013): 52–56.

Stanley, S. M., and X. Yang. Two extinction events in the Late Permian. *Science* 266 (1994): 1340–44.

Steiner, M. B., Y. Eshet, M. R. Rampino, and D. M. Schwindt. Fungal abundance spike and the Permian-Triassic boundary in the Karoo Supergroup (South Africa). *Palaeogeography, Palaeoclimatology, Palaeoecology* 194 (2003): 405–14.

Sun, Y., M. M. Joachimski, P. B. Wignall, C. Yan, Y. Chen, H. Jiang, L. Wang, and X. Lai. Lethally hot temperatures during the Early Triassic greenhouse. *Science* 338 (2012): 366–70.

Sweet, W. C., Z. Yang, J. M. Dickins, and H. Yin. *Permo-Triassic Events in the Eastern Tethys*. Cambridge: Cambridge University Press, 1992.

Twitchett, R. J., C. Looy, R. Morante, H. Visscher, and P. B. Wignall. Rapid and synchronous collapse of marine and terrestrial ecosystems during the end-Permian biotic crisis. *Geology* 29 (2001): 351–54.

Verma, H. C., C. Upadhyay, R. P. Tripathi, A. D. Shukla, and N. Bhandari. Evidence of impact at the Permian/Triassic boundary from Mossbauer spectroscopy. *Hyperfine Interactions* 141 (2002): 357–60.

Visscher, H., H. Brinkhuis, D. L. Dilcher, W. C. Elsik, Y. Eshet, C. V. Looy, M. R. Rampino, and A. Traverse. The terminal Paleozoic fungal event: Evidence of terrestrial ecosystem destabilization and collapse. *Proceedings of the National Academy of Sciences of the United States of America* 93 (1996): 2155–58.

Visscher, H., C. V. Looy, M. E. Collinson, H. Brinkhuis, J. H. A. van Konijnenburg-van Cittert, W. M. Kürschner, and M. A. Sephton. Environmental mutagenesis during the end-Permian ecological crisis. *Proceedings of the National Academy of Sciences of the United States of America* 101 (2004): 12952–56.

Wang, Y., P. M. Sadler, S. Shen, D. H. Erwin, Y. Zhang, X. Wang, W. Wang, J. L. Crowley, and C. M. Henderson. Quantifying the process and abruptness of the end-Permian mass extinction. *Paleobiology* 40 (2014): 113–29.

Wang, Z.-Q., and A.-S. Chen. Traces of arborescent lycopsids and dieback of the forest vegetation in relation to the terminal Permian mass extinction in North China. *Review of Palaeobotany and Palynology* 117 (2001): 217–43.

Ward, P. D., J. Botha, R. Buick, M. O. De Kock, D. H. Erwin, G. H. Garrison, J. L. Kirschvink, and R. Smith. Abrupt and gradual extinction among Late Permian land vertebrates in the Karoo Basin, South Africa. *Science* 307 (2005): 709–14.

Ward, P., D. R. Montgomery, and R. Smith. Altered river morphology in South Africa related to the Permian-Triassic extinction. *Science* 289 (2000): 1740–43.

Weidlich, O., W. Kiessling, and E. Flugel. Permian-Triassic boundary interval as a model for forcing marine ecosystem collapse by long-term atmospheric oxygen drop. *Geology* 31 (2003): 961–64.

Wignall, P. B., and A. Hallam. Anoxia as a cause of the Permian-Triassic mass extinction: Facies evidence from northern Italy and the western United States. *Palaeogeography, Palaeoclimatology, Palaeoecology* 93 (1992): 21–46.

Wignall, P. B., H. Kozur, and A. Hallam. On the timing of palaeoenvironmental changes at the Permo-Triassic (P/TR) boundary using conodont biostratigraphy. *Historical Biology* 12 (1996): 39–62.

Wignall, P. B., and R. J. Twitchett. Oceanic anoxia and the end Permian mass extinction. *Science* 272 (1996): 1155–58.

Xie, S., R. D. Pancost, J. Huang, P. B. Wignall, J. Yu, X. Tang, L. Chen, X. Huang, and X. Lai. Changes in the global carbon cycle occurred as two episodes during the Permian-Triassic crisis. *Geology* 35 (2007): 1083–86.

Xie, S., R. D. Pancost, X. Huang, D. Jiao, L. Lu, J. Huang, F. Yang, and R. P. Evershed. Molecular and isotopic evidence for episodic environmental change across the Permo/Triassic boundary at Meishan in South China. *Global and Planetary Change* 55 (2007): 56–65.

Xie, S., R. D. Pancost, H. Yin, H. Wang, and R. P. Evershed. Two episodes of microbial change coupled with Permo/Triassic faunal mass extinction. *Nature* 434 (2005): 494–97.

Xu, L., Y. Lin, W. Shen, L. Qi, L. Xie, and Z. Ouyang. Platinum-group elements of the Meishan Permian-Triassic boundary section: Evidence for flood basaltic volcanism. *Chemical Geology* 246 (2007): 55–64.

Yin, H., K. Zhang, J. Tong, Z. Yang, and S. Wu. The Global Stratotype Section and Point (GSSP) of the Permian-Triassic boundary. *Episodes* 24 (2001): 102–14.

Yu, J., J. Broutin, Z.-Q. Chen, X. Shi, H. Li, D. Chu, and Q. Huang. Vegetation changeover across the Permian-Triassic boundary in southwest China: Extinction, survival, recovery, and paleoclimate—A critical review. *Earth-Science Reviews* 149 (2015): 203–24.

Yu, J., Y. Peng, S. Zhang, F. Yang, Q. Zhao, and Q. Huang. Terrestrial events across the Permian-Triassic boundary along the Yunnan-Guizhou border, SW China. *Global and Planetary Change* 55 (2007): 193–208.

Zhang, H., C. Cao, X. Liu, and S. Shen. The terrestrial end-Permian mass extinction in South China. *Palaeogeography, Palaeoclimatology, Palaeoecology* 448 (2015): 108–24.

Zheng, Q. F., C. Q. Cao, and M. Y. Zhang. Sedimentary features of the Permian-Triassic boundary sequence of the Meishan section in Changxing County, Zhejiang Province. *Science China Earth Sciences* 56 (2013): 956–69.

9. Catastrophic Volcanic Eruptions and Extinctions

Alvarez, L. W., W. Alvarez, F. Asaro, and H. V. Michel. Extraterrestrial cause of Cretaceous/Tertiary extinction: Experimental results and theoretical interpretation. *Science* 208 (1980): 1095–1108.

Berner, R. A., and D. J. Beerling. Volcanic degassing necessary to produce a $CaCO_3$ undersaturated ocean at the Triassic-Jurassic boundary. *Palaeogeography, Palaeoclimatology, Palaeoecology* 244 (2007): 368–73.

Black, B. A., J.-F. Lamarque, C. A. Shields, L. T. Elkins-Tanton, and J. T. Kiehl. Acid rain and ozone depletion from pulsed Siberian Traps magmatism. *Geology* 42 (2014): 67–70.

Blackburn, T. J., P. E. Olsen, S. A. Bowring, N. M. McLean, D. V. Kent, J. Puffer, G. McHone, E. T. Rasbury, and M. Et-Touhami. Zircon U-Pb geochronology links the end-Triassic extinction with the Central Atlantic Magmatic Province. *Science* 340 (2013): 941–44.

Bond, D. P. J., and P. W. Wignall. Large igneous provinces and mass extinctions: An update. *Geological Society of America Special Papers* 505 (2014): 29–55.

Burgess, S. D., S. Bowring, and S.-Z. Shen. High-precision timeline for Earth's most severe extinction. *Proceedings of the National Academy of Sciences of the United States of America* 111 (2014): 3316–21

Caldeira, K., and M. R. Rampino. Carbon dioxide emissions from Deccan volcanism and a K/T boundary greenhouse effect. *Geophysical Research Letters* 17 (1990): 1299–302.

Chenet, A.-L., V. Courtillot, F. Fluteau, M. Gérard, X. Quidelleur, S. F. R. Khadri, K. V. Subbarao, and T. Thordarson. Determination of rapid Deccan eruptions across the Cretaceous-Tertiary boundary using paleomagnetic secular variation: 2. Constraints from analysis of eight new sections and synthesis for a 3500-m-thick composite section. *Journal of Geophysical Research: Solid Earth* 114 (2009): B06103.

Courtillot, V. E. *Evolutionary Catastrophes: The Science of Mass Extinctions.* New York: Cambridge University Press, 1999.

Courtillot, V. E., and P. R. Renne. On the ages of flood basalt events. *Comptes Rendus de l'Académie de sciences: Geoscience* 335 (2003): 113–40.

Deckart, K., G. Féraud, and H. Bertrand. Age of Jurassic continental tholeiites of French Guyana, Surinam and Guinea: Implications for the initial opening of the Central Atlantic Ocean. *Earth and Planetary Science Letters* 150 (1997): 205–20.

Drake, C. L., and Y. Herman. Did the dinosaurs die or evolve into red herrings? *Northwest Science* 62 (1988): 131–46.

Elkins-Tanton, L. T., and B. H. Hager. Giant meteoroid impacts can cause volcanism. *Earth and Planetary Science Letters* 239 (2005): 219–32.

Erba, E. Calcareous nannofossils and Mesozoic oceanic anoxic events. *Marine Micropaleontology* 52 (2004): 85–106.

Ernst, R. E., J. W. Head, R. Parfitt, E. Grosfils, and L. Wilson. Giant radiating dyke swarms on Earth and Venus. *Earth-Science Reviews* 39 (1995): 1–58.

Font, E., A. Nédélec, B. B. Ellwood, J. Mirão, and P. F. Silva. A new sedimentary benchmark for the Deccan Traps volcanism? *Geophysical Research Letters* 38 (2011): L24309.

Gevers, T. W. *The Life and Work of Dr. Alex L. du Toit.* Johannesburg: Geological Society of South Africa, 1949.

Girard, A., L. M. François, C. Dessert, S. Dupre, and Y. Godderis. Basaltic volcanism and mass extinction at the Permo-Triassic boundary: Environmental impact and climate modeling of the global carbon cycle. *Earth and Planetary Science Letters* 234 (2005): 207–21.

Hallam, A., and P. B. Wignall. *Mass Extinctions and Their Aftermath.* Oxford: Oxford University Press, 2002.

Hesselbo, S. P., S. A. Robinson, F. Surlyk, and S. Piasecki. Terrestrial and marine extinction at the Triassic-Jurassic boundary synchronized with major carbon-cycle perturbation: A link to initiation of massive volcanism? *Geology* 30 (2002): 251–54.

Hotinski, R. M., K. L. Bice, L. R. Kump, R. G. Najjar, and M. A. Arthur. Ocean stagnation and end-Permian anoxia. *Geology* 29 (2001): 7–10.

Jerram, D. A., H. H. Svensen, S. Planke, A. G. Polozov, and T. H. Torsvik. The onset of flood volcanism in the north-western part of the Siberian Traps: Explosive volcanism versus effusive lava flows. *Palaeogeography, Palaeoclimatology, Palaeoecology* 441 (2015): 38–50.

Jones, A. P., G. D. Price, N. J. Price, P. S. DeCarli, and R. A. Clegg. Impact induced melting and the development of large igneous provinces. *Earth and Planetary Science Letters* 202 (2002): 551–61.

Jourdan, F., K. Hodges, B. Sell, U. Schaltegger, M. T. D. Wingate, L. Z. Evins, U. Söderlund, P. W. Haines, D. Phillips, and T. Blenkinsop. High-precision dating of the Kalkarindji large igneous province, Australia, and synchrony with the Early-Middle Cambrian (Stage 4–5) extinction. *Geology* 42 (2014): 543–46.

Keller, G., and A. C. Kerr, eds. Volcanism, impacts, and mass extinctions: Causes and effects. *Geological Society of America Special Papers* 505 (2014).

Kelley, S. P. The geochronology of large igneous provinces, terrestrial impact craters, and their relationship to mass extinctions on Earth. *Journal of the Geological Society* 164 (2007): 923–36.

Knight, K. B., S. Nomade, P. R. Renne, A. Marzoli, H. Bertrand, and N. Youbi. The Central Atlantic Magmatic Province at the Triassic-Jurassic boundary: Paleomagnetic and ^{40}Ar/^{39}Ar evidence from Morocco for brief episodic volcanism. *Earth and Planetary Science Letters* 228 (2004): 141–60.

Kravchinsky, V. A. Paleozoic large igneous provinces of northern Eurasia: Correlation with mass extinction events. *Global and Planetary Change* 86–87 (2012): 31–36.

McHone, G. J. G. Broad-terrane Jurassic flood basalts across northeastern North America. *Geology* 24 (1996): 319–22.

McLean, D. M. Deccan Trap mantle degassing in the terminal Cretaceous marine extinction. *Cretaceous Research* 61 (1985): 235–39.

McLean, D. M. A test of terminal Mesozoic catastrophe. *Earth and Planetary Science Letters* 53 (1981): 103–108.

Officer, C. B. Extinctions, iridium, and shocked minerals associated with the Cretaceous/Tertiary transition. *Journal of Geological Education* 38 (1990): 402–25.

Officer, C. B., and C. L. Drake. The Cretaceous-Tertiary transition. *Science* 219 (1983): 1383–90.

Officer, C. B., and C. L. Drake. Terminal Cretaceous environmental events. *Science* 227 (1985): 1161–67.

Officer, C. B., A. Hallam, C. L. Drake, and J. D. Devine. Late Cretaceous and paroxysmal Cretaceous/Tertiary extinctions. *Nature* 326 (1987): 143–49.

Olsen, P. E. Giant lava flows, mass extinctions and mantle plumes. *Science* 284 (1999): 604–5.

Palfy, J., and P. L. Smith. Synchrony between Early Jurassic extinction, oceanic anoxic event, and the Karoo-Ferrar flood basalt volcanism. *Geology* 28 (2000): 747–50.

Rampino, M. R. Mass extinctions of life and catastrophic flood basalt volcanism. *Proceedings of the National Academy of Sciences of the United States of America* 107 (2010): 6555–56.

Rampino, M. R., and S. Self. Large igneous provinces and biotic extinctions. In *The Encyclopedia of Volcanoes*, 2nd ed., edited by H. Sigurdsson, B. Houghton, S. McNutt, H. Rymer, and J. Stix, 1049–58. London: Academic Press, 2015.

Rampino, M. R., and R. B. Stothers. Flood basalt volcanism during the past 250 million years. *Science* 241 (1988): 663–68.

Ravizza, G., and B. Peucker-Ehrenbrink. Chemostratigraphic evidence of Deccan volcanism from the marine Osmium isotope record. *Science* 302 (2003): 1392–95.

Renne, P. R., C. J. Sprain, M. A. Richards, S. Self, L. Vandewrkluysen, and R. Pande. State shift of Deccan volcanism at the Cretaceous-Paleogene boundary, possibly induced by impact. *Science* 350 (2015): 76–78.

Richards, M. A., R. A. Duncan, and V. E. Courtillot. Flood basalts and hot-spot tracks: Plume heads and tails. *Science* 246 (1989): 103–7.

Robinson, N., G. Ravizza, R. Coccioni, B. Peucker-Ehrenbrink, and R. Norris. A high-resolution marine osmium isotope record for the late Maastrichtian: Distinguishing the chemical fingerprints of the Deccan and KT impactor. *Earth and Planetary Science Letters* 281 (2008): 159–68.

Rocchia, R., D. Boclet, V. Courtillot, and J. J. Jaeger. A search for iridium in the Deccan Traps and intertraps. *Geophysical Research Letters* 15 (1988): 812–15.

Rothman, D. H., G. P. Fournier, K. L. French, E. J. Alm, E. A. Boyle, C. Cao, and R. E. Summons. Methanogenic burst in the end-Permian carbon cycle. *Proceedings of the National Academy of Sciences of the United States of America* 111 (2014): 5462–67.

Ruhl, M., N. R. Bonis, G.-J. Reichart, J. S. Sinninghe Damsté, and W. M. Kürschner. Atmospheric carbon injection linked to end-Triassic mass extinction. *Science* 333 (2011): 430–34.

Saunders, A., and M. Reichow. The Siberian Traps and the end-Permian mass extinction: A critical review. *Chinese Science Bulletin* 54 (2009): 20–37.

Schaller, M. F., J. D. Wright, and D. V. Kent. Atmospheric pCO_2 perturbations associated with the Central Atlantic Magmatic Province. *Science* 331 (2012): 1404–7.

Schmidt, A., K. S. Carslaw, G. W. Mann, M. Wilson, T. J. Breider, S. J. Pickering, and T. Thordarson. The impact of the 1783–1784 AD Laki eruption on global aerosol formation processes and cloud condensation nuclei. *Atmospheric Chemistry and Physics* 10 (2010): 6025–41.

Schoene, B., K. M. Samperton, M. P. Eddy, G. Keller, T. Adatte, S. A. Bowring, S. F. R. Khadri, and B. Gertsch. U-Pb geochronology of the Deccan Traps and relation to the end-Cretaceous mass extinction. *Science* 347 (2015): 182–84.

Self, S., A. Schmidt, and T. A. Mather. Emplacement characteristics, time scales, and volcanic gas release rates of continental flood basalts on Earth. *Geological Society of America Special Papers* 505 (2014): 319–35.

Self, S., M. Widdowson, T. Thordarson, and A. E. Jay. Volatile fluxes during flood basalt eruptions and potential effects on the global environment: A Deccan perspective. *Earth and Planetary Science Letters* 248 (2006): 518–31.

Sinton, C. W., and R. A. Duncan. Potential links between ocean plateau volcanism and global ocean anoxia at the Cenomanian-Turonian boundary. *Economic Geology* 92 (1997): 836–43.

Stothers, R. B. Flood basalts and extinction events. *Geophysical Research Letters* 20 (1993): 1399–402.

Sun, Y., M. M. Joachimski, P. B. Wignall, C. Yan, Y. Chen, H. Jiang, L. Wang, and X. Lai. Lethally hot temperatures during the Early Triassic greenhouse. *Science* 338 (2012): 366–70.

Svensen, H., F. Corfu, S. Polteau, Ø. Hammer, and S. Planke. Rapid magma emplacement in the Karoo Large Igneous Province. *Earth and Planetary Science Letters* 325–326 (2012): 1–9.

Svensen, H., S. Planke, L. Chevallier, A. Malthe-Sørenssen, F. Corfu, and B. Jamtveit. Hydrothermal venting of greenhouse gases triggering Early Jurassic global warming. *Earth and Planetary Science Letters* 256 (2007): 554–66.

Svensen, H., S. Planke, A. Malthe-Sørenssen, B. Jamtveit, R. Myklebust, T. R. Eidem, and S. S. Rey. Release of methane from a volcanic basin as a mechanism for initial Eocene global warming. *Nature* 429 (2004): 542–45.

Svensen, H., S. Planke, A. G. Polozov, N. Schmidbauer, F. Corfu, Y. Y. Podladchikov, and B. Jamtveit. Siberian gas venting and the end-Permian environmental crisis. *Earth and Planetary Science Letters* 277 (2009): 490–500.

Thordarson, T., and S. Self. Sulfur, chlorine, and fluorine degassing and atmospheric loading by the Roza eruption, Columbia River Basalt Group, Washington, USA. *Journal of Volcanology and Geothermal Research* 74 (1996): 49–73.

Thordarson, T., S. Self, N. Óskarsson, and T. Hulsebosch. Sulfur, chlorine, and fluorine degassing and atmospheric loading by the 1783–1784 Laki (Skaftár Fires) eruption in Iceland. *Bulletin of Volcanology* 58 (1996): 205–25.

Wignall, P. B. Large igneous provinces and mass extinctions. *Earth-Science Reviews* 53 (2001): 1–33.

Wignall, P. B. *The Worst of Times: How Life on Earth Survived Eighty Million Years of Extinctions*. Princeton, N.J.: Princeton University Press, 2015.

Williams, D. A., and R. Greeley. Assessment of antipodal-impact terrains on Mars. *Icarus* 110 (1994): 196–202.

Yang, J., P. A. Cawood, Y. Du, H. Huang, H. Huang, and P. Tao. Large igneous province and magmatic arc sourced Permian-Triassic volcanogenic sediments in China. *Sedimentary Geology* 261–262 (2012): 120–31.

Yin, H., S. Huang, K. Zhang, F. Yang, M. Ding, X. Bi, and S. Zhang. Volcanism at the Permian-Triassic boundary in South China and its effects on mass extinctions. *Acta Geologica Sinica* 2 (1989): 417–31.

Zhu, D.-C., S.-L. Chung, X.-X. Mo, Z.-D. Zhao, Y. Niu, B. Song, and Y.-H. Yang. The 132 Ma Comei-Bunbury large igneous province: Remnants identified in present-day southeastern Tibet and southwestern Australia. *Geology* 37 (2009): 581–86.

10. Ancient Glaciers or Impact-Related Deposits?

Ahmad, F. An ancient tillite in central India. *Quarterly Journal of the Geological, Mining and Metallurgical Society of India* 27 (1955): 157–61.

Amor, K., S. P. Hesselbo, D. Porcelli, S. Thackrey, and J. Parnell. A Precambrian proximal ejecta blanket from Scotland. *Geology* 36 (2008): 304–6.

Arnaud, E., and C. H. Eyles. Catastrophic mass failure of a Neoproterozoic glacially influenced continental margin, the Great Breccia, Port Askaig Formation, Scotland. *Sedimentary Geology* 151 (2002): 313–33.

Arnaud, E., and C. H. Eyles. Neoproterozoic environmental change recorded in the Port Askaig Formation, Scotland: Climate vs tectonic controls. *Sedimentary Geology* 183 (2006): 99–124.

Benn, D. I., and A. R. Prave. Subglacial and proglacial glacitectonic deformation in the Neoproterozoic Port Askaig Formation, Scotland. *Geomorphology* 75 (2006): 266–80.

Bjørlykke, K. The Eocambrian "Reusch Moraine" at Bigganjargga and the geology around Varangerfjord, northern Norway. *Norges Geologiske Untersøkelse* 251 (1967): 19–44.

Branney, M. J., and R. J. Brown. Impactoclastic density current emplacement of terrestrial meteorite-impact ejecta and the formation of dust pellets and accretionary lapilli: Evidence from Stac Fada, Scotland. *Journal of Geology* 119 (2011): 275–92.

Cahen, L. Glaciations anciennes et dérive des continents. *Annales de la Société Géologique de Belgique* 86 (1963): 19–83.

Chao, E. C. T. Mineral-produced high-pressure striae and clay-polish: Key evidence for non-ballistic transport of ejecta from the Ries crater. *Science* 194 (1976): 615–18.

Chao, E. C. T., R. Huttner, and H. Schmidt-Kaler. *Principal Exposures of the Ries Meteorite Crater in Southern Germany.* Munich: Bayerisches Geologisches Landesamt, 1978.

Chumakov, N. M. Mesozoic tilloids of the middle Volga, USSR. In *Earth's Pre-Pleistocene Glacial Record*, edited by M. H. Hambrey and W. B. Harland, 570. Cambridge: Cambridge University Press, 1981.

Crosta, A. P., and J. J. Thome Filho. Geology and impact features of the Domo de Araguainha Astrobleme, states of Mato Grosso and Goias, Brazil. Paper presented at the 31st International Geological Congress, Rio de Janeiro, Brazil, 2000.

Crowell, J. C. Climatic significance of sedimentary deposits containing dispersed megaclasts. In *Problems in Palaeoclimatology*, edited by A. E. M. Nairn, 86–89. London: Wiley, 1964.

Davison, S., and M. J. Hambrey. Indications of glaciations at the base of the Proterozoic Stoer Group (Torridonian), NW Scotland. *Journal of the Geological Society* 153 (1996): 139–49.

Edwards, M. B. Discussion of glacial or non-glacial origin for the Bigganjargga tillite, Finnmark, northern Norway. *Geological Magazine* 134 (1997): 873–76.

Eyles, C. H. Glacially- and tidally-influenced shallow marine sedimentation of the Late Precambrian Port Askaig Formation, Scotland. *Palaeogeography, Palaeoclimatology, Palaeoecology* 68 (1988): 1–25.

Eyles, C. H., N. Eyles, and A. D. Miall. Models of glaciomarine sedimentation and their application to the interpretation of ancient glacial sequences. *Palaeogeography, Palaeoclimatology, Palaeoecology* 51 (1985): 15–84.

Eyles, N. Marine debris flows: Late Precambrian tillites of the Avalonian-Cadomian orogenic belt. *Palaeogeography, Palaeoclimatology, Palaeoecology* 79 (1990): 73–98.

Eyles, N., and N. Januszczak. "Zipper-rift": A tectonic model for Neoproterozoic glaciations during the breakup of Rodinia after 750 Ma. *Earth-Science Reviews* 65 (2004): 1–73.

Hambrey, M. J., and W. B. Harland, eds. *Earth's Pre-Pleistocene Glacial Record.* Cambridge: Cambridge University Press, 1981.

Horz, F., R. Ostertag, and D. A. Rainey. Bunte Breccia of the Ries: Continuous deposits of large impact craters. *Reviews of Geophysics and Space Physics* 21 (1983): 1667–725.

Jensen, P. A., and E. Wulff-Pedersen. Glacial or non-glacial origin of the Bigganjargga tillite, Finnmark, northern Norway. *Geological Magazine* 133 (1996): 137–45.

Kim, S. B., S. K. Chough, and S. S. Chun. Bouldery deposits in the lowermost part of the Cretaceous Kyokpori Formation, SW Korea: Cohesionless debris flows and debris falls on a steep gradient delta slope. *Sedimentary Geology* 98 (1995): 97–119.

Link, P. K., J. M. G. Miller, and N. Christie-Blick. Glaciomarine facies in a continental rift environment: Neoproterozoic rocks of the western United States Cordillera. In *Earth's Glacial Record*, edited by M. Deynoux, J. M. Miller, and E. W. Domack, 29–46. Cambridge: Cambridge University Press, 1994.

Lopez-Gamundi, O. R. Thin-bedded diamictites in the glaciomarine Hoyada Verde Formation (Carboniferous), Calingasta-Uspallate Basin, western Argentina: A discussion on the emplacement conditions of subaqueous cohesive debris flows. *Sedimentary Geology* 73 (1991): 247–56.

Masaitis, V. L. Impact structures of northeastern Eurasia: The territories of Russia and adjacent countries. *Meteoritics & Planetary Science* 34 (1999): 691–711.

Mathur, S. M. The Middle Precambrian Gangau Tillite, Bijawar Group, Central India. In *Earth's Pre-Pleistocene Glacial Record*, edited by M. H. Hambrey and W. B. Harland, 428–30. Cambridge: Cambridge University Press, 1981.

Miall, A. D. Glaciomarine sedimentation in the Gowganda Formation (Huronian), Northern Ontario. *Journal of Sedimentary Research* 53 (1983): 477–91.

Middleton, G. V., and M. A. Hampton. Sediment gravity flows: Mechanisms of flow and deposition. In *Turbidites and Deep-Water Sedimentation: Short Course Notes*, edited by G. V. Middleton and A. Bouma, 1–38. Tulsa, Okla.: Society of Economic Paleontologists and Mineralogists, 1973.

Miller, J. M. G. The Proterozoic Konarock Formation, southwestern Virginia: Glaciomarine facies in a continental rift. In *Earth's Glacial Record*, edited by M. Deynoux, J. M. Miller, and E. W. Domack, 47–59. Cambridge: Cambridge University Press, 1994.

Mu, Y. Luoquan Tillite of the Sinian System in China. In *Earth's Pre-Pleistocene Glacial Record*, edited by M. H. Hambrey and W. B. Harland, 402–13. Cambridge: Cambridge University Press, 1981.

Mustard, P. S., and J. A. Donaldson. Early Proterozoic ice-proximal glaciomarine deposition: The lower Gowganda Formation at Cobalt Ontario, Canada. *Geological Society of America Bulletin* 98 (1987): 373–87.

Oberbeck, V. R., F. Horz, and T. Bunch. Impacts, tillites and the breakup of Gondwanaland: A second reply. *Journal of Geology* 102 (1994): 485–89.

Oberbeck, V. R., J. R. Marshall, and H. Aggarwal. Impacts, tillites and the breakup of Gondwanaland. *Journal of Geology* 101 (1993): 1–19.

Ocampo, A. C., K. O. Pope, and A. G. Fischer. Carbonate ejecta from the Chicxulub crater: Evidence for ablation and particle interactions under high temperatures and pressures. In *Proceedings of the 28th Lunar and Planetary Science Conference,* 1035–1036. Houston: Lunar and Planetary Institute, 1997.

Ocampo, A. C., K. O. Pope, and A. G. Fischer. Ejecta blanket deposits of the Chicxulub crater from Albion Island, Belize. *Geological Society of America Special Papers* 307 (1996): 75–88.

Parnell, J., D. Mark, A. E. Fallick, A. Boyce, and S. Thackrey. The age of the Mesoproterozoic Stoer Group sedimentary and impact deposits, NW Scotland. *Journal of the Geological Society* 168 (2011): 349–58.

Pincus, M. R., A. Cavoisie, and R. Gibbon. Preservation of Vredefort-derived shocked minerals in 300 Ma Dwyka Group tillite, South Africa. *Geological Society of America, Abstracts with Programs* 46 (2014): 706.

Pope, K. O., and A. C. Ocampo. Chicxulub high altitude ballistic ejecta from central Belize. In *Proceedings of the 31st Lunar and Planetary Science Conference,* 1419. Houston: Lunar and Planetary Institute, 2000.

Pope, K. O., A. C. Ocampo, A. G. Fischer, W. Alvarez, B. W. Fouke, C. L. Webster, F. J. Vega, J. Smit, A. E. Fritsche, and P. Claeys. Chicxulub impact ejecta from Albion Island, Belize. *Earth and Planetary Science Letters* 170 (1999): 351–64.

Pope, K. O., A. C. Ocampo, A. G. Fischer, F. Vega, D. E. Ames, D. T. King, B. Fouke, R. J. Wachtman, and G. Kleteschka. Chicxulub impact ejecta deposits in southern Quintana Roo, Mexico, and central Belize. *Geological Society of America Special Papers* 384 (2005): 171–90.

Rampino, M. R. Tillites, diamictites, and ballistic ejecta of large impacts. *Journal of Geology* 102 (1994): 439–56.

Rampino, M. R., K. Ernstson, A. G. Fischer, D. T. King Jr., A. C. Ocampo, and K. O. Pope. Characteristics of clasts in K/T debris-flow diamictites in Belize compared with other known proximal ejecta deposits. *Geological Society of America, Abstracts with Programs* 28 (1996): A-182.

Reimhold, W. U., V. von Brunn, and C. Koeberl. Are diamictites impact ejecta? No supporting evidence from South African Dwyka Group diamictites. *Journal of Geology* 105 (1997): 517–30.

Rice, A. H. N., and C.-C. Hofmann. Evidence for a glacial origin of Neoproterozoic III striations at Oaibaccannjar'ga, Finnmark, northern Norway. *Geological Magazine* 137 (2000): 355–66.

Sable, E. G., and F. Maldonado. Breccias and megabreccias, Markagunt Plateau, southwestern Utah: Origin, age, and transport directions. *U.S. Geological Survey Bulletin* 2153-H (1997): 153–76.

Schermerhorn, L. J. G. Late Precambrian mixtites: Glacial and/or nonglacial? *American Journal of Science* 274 (1974): 673–824.

Simms, M. J. The Stac Fada impact ejecta deposit and the Lairg Gravity Low: Evidence for a buried Precambrian impact crater in Scotland. *Proceedings of the Geologists' Association* 126 (2015): 742–61.

Thomas, R. J., V. von Brunn, and C. G. A. Marshall. A tectono-sedimentary model for the Dwyka Group in southern Natal, South Africa. *South African Journal of Geology* 93 (1990): 809–17.

Visser, J. N. J. The problems of recognizing ancient subaqueous debris flow deposits in glacial sequences. *Transactions of the Geological Society of South Africa* 86 (1983): 127–35.

Visser, J. N. J. Submarine debris flow deposits from the Upper Carboniferous Dwyka Tillite Formation in the Kalahari Basin, South Africa. *Sedimentology* 30 (1983): 511–23.

von Brunn, V., and D. J. C. Gold. Diamictite in the Archaean Pongola Sequence of southern Africa. *Journal of African Earth Sciences* 16 (1993): 367–74.

von Brunn, V., and C. P. Gravenor. A model for late Dwyka glaciomarine sedimentation in the eastern Karoo Basin. *Transactions of the Geological Society of South Africa* 86 (1983): 199–209.

Wang, Y., L. Songhian, G. Zhengjia, L. Weixing, and M. Guogan. Sinian tillites in China. In *Earth's Pre-Pleistocene Glacial Record*, edited by M. H. Hambrey and W. B. Harland, 386–401. Cambridge: Cambridge University Press, 1981.

Williams, G. E., and P. W. Schmidt. Origin and paleomagnetism of the Mesoproterozoic Gangau tilloid (basal Vindhyan Supergroup), central India. *Precambrian Research* 79 (1996): 307–25.

11. The Shiva Hypothesis

Alvarez, W. Toward a theory of impact crises. *Eos, Transactions of the American Geophysical Union* 67 (1986): 649–58.

Alvarez, W., and R. A. Muller. Evidence from crater ages for periodic impacts on the Earth. *Nature* 308 (1984): 718–20.

Bailer-Jones, C. A. L. Bayesian time series analysis of terrestrial impact cratering. *Monthly Notices of the Royal Astronomical Society* 416 (2011): 1163–80.

Bailer-Jones, C. A. L. The evidence for and against astronomical impacts on climate change and mass extinctions: A review. *International Journal of Astrobiology* 8 (2009): 213–39.

Chang, H.-H., and H.-K. Moon. Time-series analysis of terrestrial impact crater records. *Publications of the Astronomical Society of Japan* 57 (2005): 487–95.

Clube, S. V. M., and W. M. Napier. Giant comets and the galaxy: Implications of the terrestrial record. In *The Galaxy and the Solar System*, edited by R. Smoluchowski, J. N. Bahcall, and M. S. Matthews, 260–85. Tucson: University of Arizona Press, 1986.

Clube, S.V. M., and W. M. Napier. The role of episodic bombardment in geophysics. *Earth and Planetary Science Letters* 57 (1982): 251–62.

Davis, M., P. Hut, and R. A. Muller. Extinction of species by periodic comet showers. *Nature* 308 (1984): 715–17.

Fogg, M. J. The relevance of the background impact flux to cyclic impact/mass extinction hypotheses. *Icarus* 79 (1989): 382–95.

Grieve, R. A. F., V. L. Sharpton, A. K. Goodacre, and J. B. Garvin. A perspective on the evidence for periodic cometary impacts on Earth. *Earth and Planetary Science Letters* 76 (1985): 1–9.

Grieve, R. A. F., and E. M. Shoemaker. The record of past impacts on earth. In *Hazards Due to Asteroids and Comets*, edited by T. Gehrels, 417–62. Tucson: University of Arizona Press, 1994.

Hallam, A. The end-Triassic mass extinction event. *Geological Society of America Special Papers* 247 (1990): 577–83.

Hills, J. G. Comet showers and the steady-state infall of comets from the Oort cloud. *Astronomical Journal* 86 (1981): 1730–40.

Innanen, K. A., A. T. Patrick, and W. W. Duley. The interaction of the spiral density wave with the sun's galactic orbit. *Astrophysics and Space Science* 57 (1978): 511–15.

Lieberman, B. S., and A. L. Melott. Whilst this planet has gone cycling on: What role for periodic astronomical phenomena in large-scale patterns in the history of life? In *Earth and Life: Global Biodiversity, Extinction Intervals and Biogeographic Perturbations Through Time*, edited by J. Talent, 37–50. Dordrecht: Springer, 2012.

Matese, J. J., P. G. Whitman, K. A. Innanen, and M. J. Valtonen. Periodic modulation of the Oort cloud comets by the adiabatically changing galactic tide. *Icarus* 116 (1991): 255–68.

Matese, J., and D. Whitmire. Tidal imprint of distant Galactic matter on the Oort cloud. *Astrophysical Journal* 472 (1996): L41–43.

Matsumoto, M., and H. Kubotani. A statistical test for correlation between crater formation rate and mass extinctions. *Monthly Notices of the Royal Astronomical Society* 282 (1996): 1407–12.

Melott, A. L., and R. K. Bambach. Nemesis reconsidered. *Monthly Notices of the Royal Astronomical Society* 407 (2010): L99–102.

Perlmutter, S., and R. A. Muller. Evidence for comet storms in meteorite ages. *Icarus* 74 (1988): 369–73.

Prokoph, A., A. D. Fowler, and R. T. Patterson. Evidence for periodicity and nonlinearity in a high-resolution record of long-term evolution. *Geology* 28 (2000): 867–70.

Rampino, M. R. Role of the galaxy in periodic impacts and mass extinctions on the Earth. *Geological Society of America Special Papers* 356 (2002): 667–78.

Rampino, M. R., and K. Caldeira. Periodic impact cratering and extinction events over the last 260 million years. *Monthly Notices of the Royal Astronomical Society* 454 (2015): 3480–84.

Rampino, M. R., and B. M. Haggerty. Extraterrestrial impacts and mass extinctions of life. In *Hazards Due to Asteroids and Comets*, edited by T. Gehrels, 827–57. Tucson: University of Arizona Press, 1994.

Rampino, M. R., and B. M. Haggerty. Impact crises and mass extinctions: A working hypothesis. *Geological Society of America Special Papers* 307 (1996): 11–30.

Rampino, M. R., and B. M. Haggerty. The Shiva hypothesis: Impact crises, mass extinctions, and the galaxy. *Earth, Moon, and Planets* 72 (1996): 441–60.

Rampino, M. R., and R. B. Stothers. Geological rhythms and cometary impacts. *Science* 226 (1984): 1427–31.

Rampino, M. R., and R. B. Stothers. Geologic periodicities and the galaxy. In *The Galaxy and the Solar System*, edited by R. Smoluchowski, J. N. Bahcall, and M. S. Matthews, 241–59. Tucson: University of Arizona Press, 1986.

Rampino, M. R., and R. B. Stothers. Terrestrial mass extinctions, cometary impacts and the sun's motion perpendicular to the galactic plane. *Nature* 308 (1984): 709–12.

Randall, L. *Dark Matter and the Dinosaurs: The Astounding Interconnectedness of the Universe*. New York: HarperCollins, 2015.

Randall, L., and M. Reece. Dark matter as a trigger for periodic comet impacts. *Physical Review Letters* 112 (2014): 161301-1–5.

Raup, D. M., and J. J. Sepkoski Jr. Periodicity of extinctions in the geologic past. *Proceedings of the National Academy of Sciences of the United States of America* 81 (1984): 801–5.

Schwartz, R. D., and P. B. James. Periodic mass extinctions and the sun's oscillation about the galactic plane. *Nature* 308 (1984): 712–13.

Shoemaker, E. M. Long-term variations in the impact cratering rate on Earth. In *Meteorites: Their Flux with Time and Impact Effects*, edited by M. M. Grady, R. Hutchison, G. J. H. McCall, and D. A. Rothery, 7–10. London: Geological Society. 1998.

Shoemaker, E. M., and R. F. Wolfe. Mass extinctions, crater ages, and comet showers. In *The Galaxy and the Solar System*, edited by R. Smoluchowski, J. N. Bahcall, and M. S. Matthews, 338–86. Tucson: University of Arizona Press, 1986.

Stothers, R. B. Galactic dark matter, terrestrial impact cratering and the law of large numbers. *Monthly Notices of the Royal Astronomical Society* 300 (1998): 1098–104.

Stothers, R. B. Impact cratering at geologic stage boundaries. *Geophysical Research Letters* 20 (1993): 887–91.

Stothers, R. B. Mass extinctions and missing matter. *Nature* 311 (1984): 17.

Thaddeus, P., and G. Chanan. Cometary impacts, molecular clouds and the motion of the sun perpendicular to the galactic plane. *Nature* 314 (1985): 73–75.

Urey, H. C. Cometary collisions and geological periods. *Nature* 242 (1973): 32–33.

Whitmire, D. P., and A. A. Jackson. Are periodic mass extinctions driven by a distant solar companion? *Nature* 308 (1984): 713–15.

Yabushita, S. Periodicity in the crater formation rate and implications for astronomical modeling. *Celestial Mechanics and Dynamical Astronomy* 54 (1992): 161–78.

12. Geological Upheavals and Dark Matter

Abbas, S., and A. Abbas. Volcanogenic dark matter and mass extinctions. *Astroparticle Physics* 8 (1998): 317–20.

Chen, J., V. A. Kravchinsky, and X. Liu. The 13 million year Cenozoic pulse of the Earth. *Earth and Planetary Science Letters* 431 (2015): 256–63.

Clube, S. V. M., and W. M. Napier. Galactic dark matter and terrestrial periodicities. *Quarterly Journal of the Royal Astronomical Society* 37 (1996): 617–42.

Clube, S. V. M., and W. M. Napier. Giant comets and the galaxy: Implications of the terrestrial record. In *The Galaxy and the Solar System*, edited by R. Smoluchowski,

J. N. Bahcall, and M. S. Matthews, 260–85. Tucson: University of Arizona Press, 1986.

Courtillot, V., A. Davaille, J. Besse, and J. Stock. Three distinct types of hotspots in Earth's mantle. *Earth and Planetary Science Letters* 205 (2003): 295–308.

Courtillot, V., and P. Olson. Mantle plumes link magnetic superchrons to Phanerozoic mass depletion events. *Earth and Planetary Science Letters* 260 (2007): 495–504.

Fischer, A. G., and M. A. Arthur. Secular variations in the pelagic realm. *Society of Economic Paleontologists and MineralogistsSpecial Publication* 25 (1977): 19–50.

Freese, K. Can scalar neutrinos or massive Dirac neutrinos be the missing mass? *Physics Letters B* 167 (1986): 295–300.

Gould, A. Resonant enhancements in weakly interacting massive particle capture by the Earth. *Astrophysical Journal* 321 (1987): 571–85.

Grabau, A.W. *Principles of Stratigraphy*. New York: Seiler, 1913.

Hallam, A., and P. B. Wignall. Mass extinctions and sea level changes. *Earth-Science Reviews* 48 (1999): 217–50.

Hays, J.D., J. Imbrie, and N.J. Shackleton. Variations in the Earth's orbit: Pacemaker of the ice ages. *Science* 194 (1976): 1121–32.

Hays, J. D., and W. C. Pitman III. Lithospheric plate motion, sea level changes and climatic and ecologic consequences. *Nature* 246 (1973): 18–22.

Johnson, G. L., and J. E. Rich. A 30 million year cycle in Arctic volcanism? *Journal of Geodynamics* 6 (1986): 111–16.

King, S. D., J. P. Lowman, and C. W. Gable. Episodic tectonic plate reorganizations driven by mantle convection. *Earth and Planetary Science Letters* 203 (2002): 83–91.

Krauss, L. M., M. Srednicki, and F. Wilczek. Solar system constraints and signatures for dark-matter candidates. *Physical Review D: Particles and Fields* 33 (1986): 2079–83.

Napier, W. M. Galactic periodicity and the geological record. In *Meteorites: Their Flux with Time and Impact Effects*, edited by M. M. Grady, R. Hutchison, G. J. H. McCall, and D. A. Rothery, 19–29. London: Geological Society, 1998.

Ogawa, M. Mantle convection: A review. *Fluid Dynamics Research* 40 (2008): 379–98.

Prokoph, A., H. El Bilali, and R. Ernst. Periodicities in the emplacement of large igneous provinces through the Phanerozoic: Relations to ocean chemistry and marine biodiversity evolution. *Geoscience Frontiers* 4 (2013): 263–76.

Prokoph, A., A. D. Fowler, and R. T. Patterson. Evidence for periodicity and non-linearity in a high-resolution fossil record of long-term evolution. *Geology* 28 (2000): 867–70.

Rampino, M.R. Dark matter in the galaxy and potential cycles of impacts, mass extinctions and geological events. *Monthly Notices of the Royal Astronomical Society* 448 (2015): 1816–20.

Rampino, M. R., and K. Caldeira. Episodes of terrestrial geologic activity during the past 260 million years: A quantitative approach. *Celestial Mechanics and Dynamical Astronomy* 54 (1992): 143–59.

Rampino, M. R., and K. Caldeira. Major episodes of geologic change: Correlations, time structure and possible causes. *Earth and Planetary Science Letters* 114 (1993): 215–27.

Rampino, M. R., and A. Prokoph. Are mantle plumes periodic? *Eos, Transactions of the American Geophysical Union* 94 (2013): 113–14.

Rampino, M. R., and R. B. Stothers. Geological rhythms and cometary impacts. *Science* 226 (1984): 1427–31.

Rich, J. E., G. L. Johnson, J. E. Jones, and J. Campsie. A significant correlation between fluctuations in sea-floor spreading rates and evolutionary pulsations. *Paleoceanography* 1 (1986): 85–95.

Shaviv, N. J., A. Prokoph, and J. Veizer. Is the solar system's galactic motion imprinted in the Phanerozoic climate? *Scientific Reports* 4 (2014): 6150.

Sheridan, R. E. Pulsation tectonics as the control of long-term stratigraphic cycles. *Paleoceanography* 2 (1987): 97–118.

Stothers, R. B. Mass extinctions and missing matter. *Nature* 311 (1984): 17.

Stothers, R. B. Periodicity of Earth's magnetic reversals. *Nature* 322 (1986): 444–46.

Svensmark, H. Imprint of galactic dynamics on Earth's climate. *Astronomische Nachrichten* 327 (2006): 866–70.

Tiwari, R. K., and K. N. N. Rao. Correlated variations and periodicity of global CO_2, biological mass extinctions and extra-terrestrial bolide impacts over the past 250 million years and possible geodynamical implications. *Geofizika* 15 (1998): 103–16.

Tiwari, R. K., and K. N. N. Rao. Periodicity in marine phosphorus burial rate. *Nature* 400 (1999): 31–32.

Valentine, J. W., and E. M. Moores. Plate-tectonic regulation of faunal diversity and sea level: A model. *Nature* 228 (1970): 657–59.

Wegener, A. *The Origin of Continents and Oceans*. Translated by John Biram. New York: Dover, 1967.

Wise, K. P., and T. J. M. Schopf. Was marine faunal diversity in the Pleistocene affected by changes in sea level? *Paleobiology* 7 (1981): 394–99.

Index

Numbers in italics refer to pages on which illustrations appear, and "t" refers to pages on which tables appear.